# GM: PAINT IT RED

To Doug -
Happy reading!

# GM: PAINT IT RED

## INSIDE GENERAL MOTORS' CULTURE OF FAILURE

NICHOLAS KACHMAN
WITH ETHEL BURWELL DOWLING

MARINER PUBLISHING
BUENA VISTA, VA

1 3 5 7 9 10 8 6 4 2

Library of Congress Control Number: 2015938190
GM: Paint it Red
By Nicholas Kachman
With Ethel Burwell Dowling

p. cm.
1. Biography & Autobiography: Business
2. Business & Economics: Industries—Automobile Industry
3. Business & Economics: Workplace Culture

I. Kachman, Nicholas, 1929— II. Title.
ISBN 13: 978-0-9909653-4-3 (softcover : alk. paper)

Edited by Ethel Burwell Dowling and Judy Rogers
Cover Design by Emilie Davis, Page Design by Karen Bowen

Mariner Media, Inc.
131 West 21st Street
Buena Vista, VA 24416
Tel: 540-264-0021
www.marinermedia.com

Printed in the United States of America

This book is printed on acid-free paper meeting the requirements of the American Standard for Permanence of Paper for Printed Library Materials.

# Dedication

To my dear wife, Isabel, who never questioned my career moves, never complained about my company travel or the weekends I worked away from home, and always listened to my unending GM stories with great patience, all the while raising our five close-knit children and making her own career as a wonderful schoolteacher. I really owe my good life to her and her alone.

# Contents

# Acknowledgments

To what or to whom do I give credit for the insights in this book and a career full of learning and excitement? I did not begin with the knowledge or understanding that I have now. The answer lies in the many mentors I had along the way. I am thankful for the opportunity to befriend some of the very best people in the field. I received support and nurturing from so many intelligent people. They both corrected and shaped my thinking. Naming these individuals brings back wonderful memories and a sense of pride in having known them.

Therefore a debt of gratitude…

- To David G. Hawkins, in appreciation of his dedication to the enactment of the Clean Air Act and its amendments.
- To my dearest friend, the late Jack Cragen, a rare intellect in GM, and an extraordinary engineer and family man. I could always count on Jack's willingness to discuss any issue in an atmosphere of complete frankness and with a shared enthusiasm for innovative ideas.
- To Gabriel Steel Company owner, Fred Russel, industrial genius, and Al Schaub, plant manager, for guiding a young engineer through a complete modernization of their factory, and demonstrating what good management is and what it can accomplish.
- To John Garsh, former research scientist for American Air Filter Co., for taking me under his wing into the field of air filtration, and through the development of a new high efficiency bag filter. John was the first of many outsiders serving as an educator, who made my career more successful.

- To Don Hardie, chief engineer of the Mechanical Group at Albert Kahn Architect-Engineer Associates, for being a mentor and advancing my career as a knowledgeable building construction mechanical engineer. GM Assembly Plants are air-conditioned today, based on techniques that Don, with a great deal of patience, taught me.
- To Howard Baldwin, director of Corporate Plant Engineering, a true visionary and progressive thinker, who heard of my progressive work at Chevrolet, and assigned to me the task of developing systems to address environmental problems years ahead of other companies. His predictions in the early 1960s of future energy crises and major landfill problems were almost too forward thinking for management, but he really made an impression on me.
- To Gabe Tiberio, a special debt of gratitude for the opportunity to institute and manage all of GM's plant pollution control programs during those first exciting years of EPA regulatory control.
- To Leonard Wocholski, the late physics professor at the General Motors Institute, for spending many days with me conducting factory noise, heat, and pollution measurements, leading to our development of innovative solutions. This remarkable teacher never used "rules of thumb," but always applied the classical engineering approach of defining the problem, measuring its effects, and re-measuring after a solution was in place, to determine whether engineering assumptions were correct.
- To Ken Robinson, a nationally-known industrial ventilation expert, for taking me to many different manufacturing plants to measure the performance of every type of air supply and exhaust system in practice. First-hand observation of many machining and fabrication operations was invaluable in expanding my base of manufacturing knowledge.
- To Walter C. Barber who, as director of the EPA's Office of Air Quality Planning and Standards, honored

my request for relief from federal standards, allowing a closed small truck facility to operate for a few years, thus creating thousands of needed jobs. This episode, which taught me to be honest and forthright in approaching agency officials, served the company well throughout my career.

- To Dave Eden, director of the Environmental Staff, a General Motors outsider who managed from a distance, but knew more about our employees and programs than other executives. In my mind, what made him special was his real concern for the health and welfare of current and former employees in need. I'm the only one aware of his efforts to support a number of individuals. Eden was extremely smart in hiring the country's most outstanding ex-government officials as consultants, making my work experience exciting and fulfilling.
- To George T. Wolff, PhD, former GM Atmospheric Scientist, for his outstanding air quality research projects, which I used to establish my program priorities and to allocate resources. George articulated the character quality, possessed by all the aforementioned men, which set them apart from others in their field. After retiring from GM, and upon receiving an award, Wolff said, "I'm not retired. I'm available for research. I can't stop, for it's in my blood." That passion for one's work was largely missing in the old GM, but will be needed in order for the new GM to be successful.
- To Raymond Hoste, a most remarkable fellow and the best engineering mechanic GM Engineering Staff ever had. He would repair vehicles at his home for GM vice presidents, for family members, and friends, for neighbors, and postmen, for the richest and the poorest, regardless of race, with the same quality of workmanship for all. He was a Christian who practiced his religion through his work every day. He was like a second father to me, teaching me auto mechanics and telling many fascinating product-related stories, many of which are included in

this memoir. Every visit with Ray was a memorable event and a learning experience. I wasn't alone in feeling this way about Ray. Over 300 attended his funeral—standing room only!

And I am particularly indebted to the following three men who helped me with material for this book:

- To Douglas A. Frank, a former GM EAS staff member who became the leading sales engineer for GM FANUC, the company that supplied the auto industry with robotic paint systems.
- To Dave Aldorfer, my closest engineering associate, who believed in the Paint Plan, and did his best to implement it, defend it, and excuse the plants as they failed to comply. He developed most of the technical documents sent to the state and federal agencies, and to the courts. His work ethic and the quality of his work were admired by all, but for a Plan sure to fail. Dave's written views on the Paint Plan's implementation have proven to be of great value.
- And especially to John Calcagni, retired chief of the EPA Office of Air Quality Planning and Standards. I am indebted to him, with his vast personal knowledge of VOC (volatile organic compound) controls, for the time he spent responding to my letters with his own recollections of the development of the Paint Plan.

## Without My Editor and Publisher, There Would Be No Book

My editor, Ethel Dowling, took four different treatises on the GM Paint Plan, plus dozens of handwritten stories on related subjects, and arranged them by subject matter and topic. After reading and rereading these documents, researching the subject matter and asking the right questions, she determined that we had a book that would be of interest to many. Her confidence in my content was buoyed by statements made by her late father in regard to Chrysler management. She quickly made a book outline that would attract, educate, and hold the interest of readers. For me it has been a life-fulfilling dream to have such a skillful editor to take on my story at this stage of life.

Dealing with Ethel has been an enjoyable learning experience. Smart as can be, steady as a rock, and with great integrity, she has thoughtfully considered my words and given me peace of mind by ensuring that the story is not unjust to anyone. After careful deliberation, we chose an established professional publisher with an impressive track record. Judy Rogers and Karen Bowen of Mariner Media have displayed outstanding professionalism in meticulously making corrections. And the president, Andy Wolfe, deserves special mention because of his willingness to lose sleep in taking on this unknown author who speaks candidly about the flaws of a corporate giant.

I'm confident my editor and publisher have made the very most of whatever I've given them, and if the book is well received, it will be because of what they did.

# Editor's Preface

My family moved into a home next door to the Kachmans when I was all of three years old. My father, an engineer in management with Chrysler, and Nick, with GM, swapped many company stories across the cedar picket fence that separated our driveway from the Kachman's backyard. Nick recalls working on his family vehicles with my father looking over his shoulder advising. Both men took pride and pleasure in maintaining their own vehicles.

My father, now deceased, with degrees from Yale and the Darden School at the University of Virginia, had tremendous respect for Nick. He valued Nick's insights and broad professional knowledge as well as his dry sense of humor. Though their backgrounds may have varied, they were both engineers who were unequivocally devoted to their families and loyal to their respective corporations, even during periods when there was much to critique. As the flagships of the American auto industry faltered for the first time in the '70s and '80s, there were astute insights that were shared across that fence and frustrations that were politely vented, but never a thought of jumping ship, only a hope that improvements were coming. That same hope is what I have observed to motivate Nick in his composition of this book.

Without a thorough reading, Nick's book might be assumed to be vindictive or bitter, but that is not in his character. Instead he is merely a passionate truth teller, incredulous that so few have spoken out about the problem with the "emperor's new clothes."

As I worked to shape his writings of the past five years, some typed by family and some scrawled on legal pads, into a coherent story line, I was impressed by his rise from Ford Trade School toolmaker to GM executive; I was intrigued by his insider vignettes; and I was struck repeatedly with the larger implications of his observations.

His is a story of personal success in the midst of corporate failures. It contains lessons for every employee about integrity, assertiveness, and perseverance through opposition. It contains warnings about the hazards in business of short-sighted self-preservation and the failure to create a corporate environment that values the essential elements for long-term success. Isn't long-term success what every business should seek?

In crafting his story, I have attempted to set the stage—give the reader some insight into Nick's career path and GM's history and culture—then bring them up to speed on the topics of pollution and automotive paint. All of this lays a foundation for understanding the key role that the failures of what Nick refers to as the "Paint Plan" played in the dramatic fall of an iconic American corporation.

Much of the story is rooted in the '70s and '80s, but as this book goes to press, the headlines indicate that little has changed at GM. The insights in the pages to follow are relevant and needed as much today as they were decades before.

It has been my privilege to make Nick's stories and insights accessible to all. Happy reading and learning.

—Ethel Burwell Dowling

# 1

# The Report

I DROVE THROUGH THE POURING RAIN THAT NIGHT IN 1980, STRAINING to see through the windshield of my teenager's Vega as the wipers did their best to help me see the street signs on the curvy township road. It was after 10 p.m. but I was looking for the home of Joseph Godfrey, the vice president of the General Motors Assembly Division (GMAD), so that he could have a copy, the night before, of the report that I would give to upper management in the morning. I knew that the numbers in my report would expose the failure and outrageous expense of the plan that GMAD had implemented under his dictatorial leadership. Included in my presentation were disclosures of the lost court cases, community suits and the plants' status as non-compliant, all as a result of Godfrey's Paint Plan.

These were far from ideal conditions for a trip through Detroit to the Long Lake area of Birmingham on the far north side of town, but I knew it was the right thing to do. In my years at GM, I had developed a reputation for speaking unwanted truth, but my motives were never vindictive, just honest. I left the envelope nestled into the decorative horse drawn carriage under his front porch lights. The next morning, at the Environmental Review Board meeting, he sat quietly as I made my report and people like Vice President Bob Stempel, who should have known sooner, finally grasped the magnitude of the problem, and raged that they had not been told. Stempel, manager over all the BOC (Buick, Oldsmobile, and Cadillac) plants, was taken by surprise. Embarrassed and disturbed, he called his people out of the meeting asking why he hadn't been informed. I had told Bob Phillips, his engineer, to do so but up to this point I had been silenced by the GMAD central office.

It was not long after this that Joseph Godfrey retired. The Paint Plan however, continued to wreak financial havoc as Roger Smith, who had signed off on the Plan as the GM Chief Financial Officer, moved that same year into the position of CEO.

Even before the recent concerns related to unaddressed safety issues, many have asked what could possibly push what was once the most profitable company in the world to the verge of bankruptcy. What forces combined in 2008 to bring General Motors Corporation begging for help from the federal government? Because of its long history as the biggest and most profitable company in the world, children were told that a job with GM would guarantee one a long career with wonderful benefits for life. It was a company you could stay with, a company that you could count on for stability. But the unbelievable happened and the shockwaves were felt throughout the auto industry and beyond. The bankruptcy adversely affected the lives of millions of people—thousands of families, hundreds of subsidiary companies and numerous cities and communities far beyond Detroit. It is not surprising then that much ink has already been spilled to suggest answers to this question. Some have pointed to healthcare costs or product offerings. Other analysts have cited union demands or oppressive government regulations. But from 1957 to 1993 I was there on the inside, attending key meetings, observing management decisions, sometimes advising but always watching it all unfold as I worked in my position as corporate environmental engineer. I saw first-hand the very sickness in the culture that plagues GM to this day.

There is a side to the bankruptcy story that has never been reported or revealed to the public and it involves the debacle called "The Paint Plan." In her insightful book *Rude Awakening*, Maryann Keller reports a gloomy statement by Executive Vice President F. Alan Smith. "'From 1980 through 1985,' he told the group, 'GM spent $45 billion in capital investment, yet increased its worldwide market share by only one percentage point, to 22%.'" To put this in perspective he goes on: "'For the same amount of money,' he told the GM mangers, 'we could buy Toyota and Nissan outright, instantly increasing market share to 40%.'"[1] Why was *$45 billion* spent with so little return on investment? Does this sound like a path to profitability? What Keller doesn't reveal is that most of that $45 billion was tied to the mushrooming expenses of the Paint Plan. With the original proposal giving cost estimates in 1976 of $70 million to overhaul the paint shop at each of the 31 GM plants, the actual costs were coming in ten times higher per plant (i.e. $700 million) with additional millions going to related legal fees and EPA fines. In addition, as a component of the Plan, three entirely new plants had been not only built, but in reality expensively overbuilt.

This is the side of the story that hasn't been told publicly despite the fact that, as my final chapter will show, its repercussions continue to be a financial drag on the corporation to this day. After my condemning report in 1980, the behemoth that is General Motors lumbered on toward bankruptcy, trapped in the Paint Plan because of the commitments that the Plan made to the EPA to reduce air pollution by switching to a yet undeveloped paint formula, and because Chairman of the Board Roger Smith had staked his pride and reputation on the Plan's ability to "leapfrog" GM ahead of its competition.

Eight years later, in 1988, I had a chance to raise the issue directly to F. Alan Smith, executive vice president. In the morning, reports were given to Smith by Director of Automotive Safety Robert Rogers and Director of Automotive Emission Control Samuel A. Leonard. Each presented their most prominent issues, demonstrating their importance to the corporation so as to impress Smith and not bore the man. At lunch the directors both reported with some astonishment that Smith had closed his eyes, put his head down, as if asleep, and listened (they think) without asking a question of either director. What chance did I have in presenting my talk after a luncheon meal?

At 1:30 p.m., I got my chance to make a report on behalf of the Plant Environmental Activities Staff. After a few minutes, Smith already had his head down, eyes closed and appeared to be asleep. In my mind, the Paint Plan shenanigans were the biggest plant-related headaches GM had, so I took the gamble and said, "We would never do that again!"

To everyone's surprise, Smith slowly raised his head, opened his eyes wide, and with the fierce look of a Norwegian warrior said, "You bet we won't."

Shamefully, I was pleased. A response like that from a strictly financial man clearly indicated that Godfrey's folly was finally receiving bad reviews at the top. But why had it taken so long? To answer this question you have to understand GM.

In 2009, Steven Rattner, the government bankruptcy expert who directed the GM case, summed up his thoughts this way: "In the case of GM, the overarching question is whether, without an infusion of new blood, its management team can implement the massive cultural change that is essential."[2] When I heard this statement I knew that many outside the company, and even some within, would wonder about the management and culture issues to which he referred; but I had observed these issues first hand and been troubled by them for years. I knew he was right. I knew it was time for my specific angle on the inside story to be told and I knew that there would be a benefit for the public and the company. Many of the issues in management and in the corporate culture of GM are not specific to the auto industry. In fact, to the contrary, they have infected and poisoned many a company, many an organization, and even the government.

The blinding nature of pride and the protective nature of self-interest can bring chaos and destruction to any institution of any size. Most of the flaws in leadership and management that I observed at General Motors had their roots in these two human conditions, pride and self-interest, and the fruit that they bore was bankruptcy.

But to effect positive change, the problem must first be thoroughly understood and questions need to be answered. To give a broad stroke summary, the problem originated at the highest levels, had its core in the remnant of Fisher Body people that ran North American Operations through GMAD, infected every level of operations, was not available for critique or change, and negatively impacted every GM employee, customer, stock holder, and eventually every tax paying citizen. But that is only scratching the surface. A thorough understanding of these issues in GM's history is essential to setting GM on a profitable path for the future and these observations of GM provide a learning opportunity for all.

The following pages will reveal the detailed answers to the questions of "Who? What? When? Where? and How?" At the core of the story is "The Paint Plan." Conceived and implemented in 1978, it was the ultimate expression of pride and self-interest, and is the epitome of every wrong practice that brought down GM. To my surprise, no newspaper, book, or automotive expert has ever mentioned this ridiculous and gigantic environmental program, which GM willingly undertook. It involved a commitment to the EPA to meet certain air pollution standards in a 10-year time frame at 37 GM manufacturing facilities. Serious financial penalties could be imposed for non-compliance once the commitment was made. In theory, this plan to upgrade every GM auto paint shop would leapfrog GM past our hated competitors, Ford and Chrysler. We would force them into a scramble to meet the higher anti-pollution standards that we would set and simultaneously we would be setting new standards in auto paint quality and application technologies. But from its inception, the Plan was seriously flawed. Both the conception and the implementation of the Plan defied all good business and engineering principles, yet pride and self-protection perhaps blinded corporate leadership to this truth. The billions of corporate dollars that it wasted have never been brought to light and it is time for its foolishness to be exposed.

The facts and figures contained in the report that I left for Godfrey on that rainy night only confirmed the concerns that I had from the beginning. The visual image of David Skivens and the Fisher Body people marching into our Environmental Activities Staff offices in 1976 is clear in my mind. They confidently announced to me that my staff would aid them in justifying to the EPA a five-year extension to meet the paint emission regulations. The leaders from Fisher Body, including Godfrey, assumed that the rest of GM was at their disposal and ready to do their bidding without question.

What would become known as "the yellow book" offered the federal agency the opportunity to set more restrictive standards on the entire auto industry. Normally one to embrace eco-friendly policies, I balked at this proposal. Quickly I saw that the Plan was doomed to failure. It made no mention of possible plant closures. It contained no indication that some older plants could never be converted because their low ceilings would not accommodate the new technology. There were no cost estimates for new plant construction to replace the old plants. It failed to consider any form of coordination with the corporate product plan and schedule. It failed to consider the alarming lack of technological readiness in the area of paints and paint application systems, and finally it failed to consider the possibility (and eventual reality) of an economic downturn. All of these factors should have been points of discussion and data gathering before the approval of a commitment to the EPA with such large-scale expenditures.

By 1976, I had already had 10 years of experience working with the GM Manufacturing Staff Department of Paint Research and Development. I understood paint. I was fully versed in the environmental problems that can come from paint emissions as brought to our attention in Los Angeles, and had studied the available options to address these concerns. Dating back to 1968, Bob Phillips of Manufacturing Staff and I had conducted the first GM emission exhaust stack tests in Tarrytown, New York, and on the Flint, Michigan, plant topcoat color booth and ovens. Since my career path had caused me to be well informed on these topics, I assumed that management would want to draw upon my in-house experience. I opposed their plan with confidence. With data to support me, I felt certain that my position would prevail—if, and that was a big "if"—I could get *past* the Fisher Body management to disclose the facts to upper management.

Knowing that the Paint Plan was close to approval, I quickly drafted a proposal, outlining an approach that could meet the law, but most importantly, was actually feasible. I presented it to the Fisher Group, but within three or four minutes they had walked out. A few days later I learned that my director, Bill Johnson, and my new vice president, David Potter, had approved the Paint Plan and without my knowledge had scheduled a meeting in Chicago to present it to the EPA. Neither man had any assembly plant experience or knew much about the painting of our vehicles, but Potter was a "yes" man, and if GM President Pete Estes approved the plan, Potter would sell it. Neither man asked me, their manager over corporate air pollution control, for my opinion on the Plan.

Whatever happened to "think, consider alternatives, debate issues" before undertaking one of the company's largest rebuilding projects of its history? There is no doubt in my mind that the GM president at that time, Pete Estes, and future GM president, James McDonald, both believed the Fisher Plan was their only choice, and with blind faith in Fisher Body's Joseph Godfrey, who believed he could gain legendary status by catapulting GM into the future in just 10 years, they approved the Plan. This was gross mismanagement. Was there no one in management to question the wisdom of this Plan or was the corporate culture already too sick?

Many times in my retirement years, I have pondered these questions and discussed them at length with family and friends. As I read books and articles about GM, I find that I am not alone in my concerns or in my observations. Recent safety recalls and plant and paint shop closures have only served to confirm my fears, but no one has yet revealed this angle of the story and the lessons it holds for the future. What amazes me is the insider observation seat that my career path afforded me. While many respected and well-heeled journalists have provided their readers with insights based on extensive research and interviews, I was observing the inner workings of GM every day and reporting to those in the most influential positions. When I began at GM, I never anticipated the level of responsibility to which I would rise, and when I look back now, it boggles my mind. My beginnings were much more humble.

# 2

# Thinking Differently

As I contact them, many of my retired co-workers have shelved their GM days and have no interest in returning to those memories. But 25 years later the untold story pecks at my insides like a chick wanting out of its shell. I read the headlines in the Detroit newspapers and I see that the issues facing GM today have their roots in the Paint Plan years. I find it fascinating to see how a corporate culture can foster smart decision-making, or in the case of GM, blind people into marching lock step with the group around them, never questioning the wisdom of the ideas and plans that are being set forth. My wife fears my writings will end my GM pension, but both then and now I have never lost confidence nor felt insecure standing alone.

Time has proven that my observations were accurate and my concerns were valid. Am I smarter? I think not! So why do I think differently? Perhaps it was my humble beginnings, assisting in the family business, and working as a tool and die maker. Or perhaps my character was forged by my early experiences working for smaller companies, engaging in various engineering disciplines. I enjoy solving problems. I like standing back and looking at the big picture. At the end of the day, if I had done something to improve the profitability of my corporation, or the health of our environment, or even the life of an individual, I found satisfaction in that. I still do. Len Sharla, a GM attorney, kindly said he never met a man who got to the bottom line of a problem more quickly than I did. Whatever the reason, what I do know is that from my first years at GM to this day, I think differently than many at GM. But maybe it wasn't me that was "different," maybe it was GM.

## Humble Beginnings

The first great influences on my life were my parents. My dad, a gentle man, devoted his whole being to my mother. She was a natural-born businesswoman who spent most of her working days running her father's food and clothing business, and later, the family's nightclub business. My dad built his own home and was a metal finisher at the Chrysler Dodge plant. He cared about his fellow workers and was instrumental in the planning of a strike which resulted in provision of a decent daily wage. Both my father and I experienced unsatisfactory working conditions in manufacturing plants. Later, these experiences proved to be the impetus for my interest in improving the environment around each worker.

Through my three brothers, I was exposed to many types of manufacturing and engineering. Before I was 12, my brothers and I worked for our Uncle Steve, making thousands of wooden fish baits—manufacturing at a simple but profitable level. As an adult, my brother, Robert, owns several patents, has designed and manufactured many products, and is a servo-valve systems designer. My brother, Donald, has a Master's Degree in Vibrations and Dynamics, worked on the Chrysler Redstone project, and later in the aerospace industry. He cannot talk about most of his work, as it is still classified by the military. My youngest brother, James, spent a few years in the Army moving ballistic missiles between countries and later joined General Motors, where he served in the transmission engineering test car garage for 33 years.

At the age of 12, I became a wage earner. I delivered 25-pound blocks of ice and bottled milk with the cream on top, from a horse-drawn carriage to wooden ice boxes that sat on the outside porch of each home. The horse knew every customer and would move slowly to the next house with the word "go."

In February of 1943, at the age of 13, I enrolled in the Ford Trade School and began my education as a toolmaker. I woke up at 5:00 a.m. to catch a city bus on the east side of Detroit. Weighing only 98 pounds and dressed in a surplus navy coat, I rode the bus through Hamtramck into Dearborn. While it was still dark outside, I transferred to a Wyoming bus that took me to the famous Gate Four (union leaders were badly beaten on this overpass) at Ford's

River Rouge plant complex. There were two extremely large open-hearth cupolas next to the front gate that roared loudly and lit up the morning sky. It was enough to scare the hell out of anyone, let alone a 13-year-old.

My first workstation was next to the passenger car assembly line, about 25 feet from the end. I worked in what, until its 2005 closure, was the oldest assembly plant in the United States. Upon graduation from trade school, I was skilled in operating several lathes, grinders, and a jig-boring machine (see photo). I could rebuild a stick shift and rewind a complete electric motor. Many of my classmates never went on to work in the manufacturing industry but after graduating in 1946, I continued Ford's apprenticeship journeyman program for another two years.

Leaving Ford at the age of 18, I went to work for Beaver Tool and Engineering Company, machining ball and screw mechanisms for large aircraft landing gears and wing flaps. When I left to attend college part-time, the company president said I could have my job back at any time, but I was preparing for bigger things. I didn't just want to know "how," I was ready to learn "why."

When I began my quest for an engineering degree at Wayne State University, I was already married to my lifetime sweetheart and

*Nick at age 13 operating a lathe, turning and threading parts, in his Ford Trade School Days. This picture, from 1942, surfaced at a class reunion.*

Credit: Unknown Classmate

worked while taking classes. Initially, I worked for a small tool and die shop making dies and parts for military tanks and worked in the design of the Chrysler Redstone Missile.

My parents also allowed me, during the first two years of college and in between jobs, to work for them as a bartender at their business. This exposure to all types of people added an important dimension to my technical education. Some of the customers made the job enjoyable with their interesting stories and friendly conversations, but I often dealt with some of the roughest people I was ever to meet. One of our customers was a weightlifter who had won five Olympic medals, including a gold, while another was head of the Detroit Eastside gang. It was not uncommon for a bloody fight to break out in our bar at least once a month. Believe it or not, learning to calmly handle these characters served me well in my conflicts within General Motors.

## Industrial Engineer: A Learning Opportunity

In 1953, while taking a time and motion study class in night school, the professor asked if any student was interested in a full-time job. I eagerly raised my hand, accepting what turned out to be the

*The author at Beaver Tool and Engineering Company in 1948, where he operated a slotter and radial drill in the manufacturing of mechanisms for large airplanes.*

Credit: Family Photo Collection

opportunity of a lifetime. After a series of personnel tests, I was hired by Mr. Fred Russel, the new owner of the Gabriel Steel Company. From age 24 to 28, I learned the job of an industrial engineer.

Fred Russel, an industrial genius and a respected lecturer at UCLA on continuous process improvement and manufacturing excellence, had been the executive author of the highly acclaimed operating procedures at the Douglas Aircraft Company. Because of his business reputation, the banks offered him the opportunity to take over the bankrupt Weiser Lock Company, which he proceeded to transform into the second largest lock company on the North American continent. His goal was to do the same for the failing Gabriel Steel Company and despite my inexperience I was to be his process and industrial engineer. At that time, Gabriel Steel specialized in the manufacturing of casement window frames, basement supports, and other home hardware items.

The first thing Mr. Russel had me do was to make a layout showing every machine and process line in the steel plant. Looking down at the completed layout, Mr. Russel asked, "What is wrong with the layout of the plant?" Not waiting for an answer, he said the materials flowed back and forth and not from one end of the plant to the other. He directed me to rearrange the process and improve or eliminate each work station. How many young engineers get an opportunity to rearrange an entire factory under such great supervision? He taught me how to look at a situation, recognize problems and then find solutions that really worked. This way of thinking became part of my DNA. To me, every day was exciting at Gabriel Steel. I never tired of improving the process that allowed raw materials to enter one end of the plant and the completed product to exit at the other end.

When it came time to implement an important plant conversion, Mr. Russel brought in Joe Witt, his process guru. Again, it was a privilege so early in my career to work under and observe such a man. Mr. Witt, a former top U.S. Government manufacturing expert during WWII, had 100 skilled machine re-builders constantly improving the lock manufacturing line. Both men believed in the value of empowering the line worker to maintain equipment quality and implement improvement. Russel's California Weiser plant had only a few janitors; each worker cleaned his own machine, whereas in General

Motors assembly plants that ratio is upside down: there are hundreds of maintenance workers but only a few skilled machinists. GMAD does all their process engineering from a distance at the Fisher Body Center office (now GMAD), curtailing innovations at the plant level.

Russel spent most of his time at Weiser Lock in California, and left his brother-in-law, Al Schaub, in charge of the plant in Detroit. Schaub gave me encouragement and support in redesigning the plant. As a trivia side note, Schaub had once broken the arm of movie star John Wayne when they were both in football practice at University of Southern California.

About every six months, Russel would fly out from California to tour the plant. I learned during these visits that Russel's expertise extended beyond the engineering side of manufacturing; he also knew how to deal with people—the workers and skilled tradesmen. He mentored me in this area as well. I recall the time we were walking on the factory floor making plans for a manufacturing equipment upgrade. I stopped and explained to a group of process engineers that the 300 unit per hour rate they were now achieving would soon be bumped up to 3,000! I envisioned that the employees would share my enthusiasm for the latest technology, but as we moved on Russel asked me if I had noticed how their faces fell with that announcement. He pointed out that the manner in which I delivered the news had devalued the current efforts of the group by making them sound woefully inadequate. His point was that most employees take pride in what they do and what they make and should be treated with respect and as members of our team. He gave me several helpful corrections and lessons along these lines.

At some point in his visit, Russel would hold a dinner meeting for me and Al Schaub at the Golden Door on Jefferson Avenue. "How many people can I let go?" he would ask. Typically, I would respond by saying 31 or, if everything went as expected, possibly 33. His answer was always the same: "I will post a notice to the union tomorrow that 33 must go." The plant had started with about 300 employees, but under Russel's leadership, Gabriel Steel was producing twice the volume of product with less than 100 employees. The turnaround of the once failing company was complete and with union demands mounting, Russel decided to sell. Looking back, I see it as one of the most remarkable experiences of my life.

After the sale of the plant, Russel told me to look for a new job and he would pay me for as long as it took to find employment. By the next Monday (May 8, 1957), despite the worst recession since WWII, General Motors Chevrolet Engineering Center Budget Department hired me to implement a preventative maintenance program for the center. Six months later, Mr. Russel paid for my wife, Isabel, and me to visit Weiser Lock in the Watts area outside Los Angeles, California, and offered me a position to work with Joe Witt. I was sorely tempted, but Los Angeles was severely polluted and General Motors was a new venture, so I didn't accept the offer and moved forward with my career with General Motors. I didn't doubt my decision to stay with General Motors. I have no regrets. It led to a great career for me and had many benefits for my family. I just looked forward to what would be a bright, interesting, and stimulating future.

## Early Years at General Motors (1957–1962)

I began my General Motors career in the Chevrolet Engineering Center. My first assignment was to complete a preventative maintenance program initiated by persons having little knowledge of how a machine shop operates. The program dictated that every electric motor was to be greased or oiled every three months and every machine belt inspected and tightened every six months without regard for the use of the machine. I recognized the flaws in the plan immediately and because of my lack of interest in pursuing an illogical program, I was transferred from the budget department to plant engineering within two years.

While working as a plant engineer in the Engineering Center I was privy to many cutting-edge projects. It was during this time that my interest turned to ventilation and air-conditioning systems. Years later, because of my interest and experience, I became the ventilation expert for the Chevrolet Motor Division. One of my proudest achievements was proving to the reluctant GMAD chief, Joseph Godfrey, that air-conditioning for the hourly employees would actually save GM money over ventilation with outside air. My interest in pollution problems also first surfaced during this time period when it was noted that electrically charged small black particles appeared on our drafting boards. We traced these particles to the Chrysler tank arsenal plant's electrostatic filters on their powerhouse and corrected the problem.

In those days, GM was making a lot of money and had over 50% of the market, so we worked Saturday and most Sundays. When the head of the mechanical section went on vacation, I took over and got into trouble. When the supervisor came back, Roy, one of my draftsmen, ran to him and said, "Nick's got all of the work out," meaning there was no reason to work overtime. But the overtime work didn't stop. The workers' desire to make money took precedence over any corporate desire for efficiency. The thinking ran counter to all I had learned from Mr. Russel at Gabriel Steel.

A few months later, the plant engineering department and maintenance department crew threw a huge going away party for me. Management, in order to "teach me a lesson," was transferring me to Chevrolet Central Office (CCO) to work for Al Black. Four years later, I was asked to come back to the Chevrolet Engineering Center, but I was learning too much to accept the offer. I was ready for new challenges. Working for Al Black was not the punishment they expected. My attention now turned from the process inside the plant to the physical plant itself.

## Mechanical Engineer at CCO (1962–1967)

The Chevrolet Central Office had 30 engineers on staff. Some were civil, some electrical, and others mechanical engineers. The Chevrolet division had more plants than all the other divisions combined. All the building projects at the plants were assigned to a CCO group of skilled engineers. For years I have been extolling the value of my foundational training at the Ford Trade School, and the thrill of my learning opportunities at Gabriel Steel. But the greatest benefit of my years at the CCO was the wonderful education I gained from working with such experts: a knowledgeable group of GM engineers and the employees of the finest architectural engineering firms in Detroit. The Chevrolet Central Office engineers analyzed appropriation requests for plant projects for the benefit of Chevrolet upper management, and as the in-house experts, we worked closely with architectural and engineering firms in the construction of new plants or add-ons to existing Chevrolet Division plants. In the 1960s, Chevrolet had more assembly plants and manufacturing plants than the rest of the GM divisions combined. As one of the 25–30 engineers in the CCO, I worked in the mechanical section dealing with mechanical piping, heat and ventilation, and powerhouses. All of my successes in my GM career can be credited to those in management who provided the resources and the freedom to learn and solve problems. In the CCO our work was valued. Manufacturing Staff, Research Staff, and the General Motors Institute (GMI) were all available to us as resources. We initiated innovative projects in the plants with a free hand. Those were unbelievable and fascinating days for any engineer. My later frustrations with the management of GMAD cannot overshadow the privilege of the opportunities I had in my years at CCO.

Don Hardie, of Albert Kahn's industrial design company, mentored and taught me most of what I know regarding air-conditioning and ventilation systems. Jack Cragen, CCO project manager, and Gus Pappas, electrical section supervisor, allowed any question to be asked and didn't mind debating any issue. This free flow of analysis guided my intellectual development. To quote my dear friend, Jack Cragen: "Being a visionary and the creator of new ideas are characteristics of a great man." Such was the atmosphere at that time in the Chevrolet Central Office.

Ken Robinson of the GM Industrial Hygiene Department, our country's leading ventilation expert, and I visited and evaluated almost every GM plant. And with my growing expertise in this field, I was invited to speak at Michigan State University's Annual Ventilation Conference.

In 1966, I was the Central Office mechanical engineer in the building of the new Lordstown Assembly Plant in Ohio. I take pride in the fact that for the first time, the ventilation units were held to just three standard sizes, resulting in tremendous cost savings.

During this time I worked simultaneously as the mechanical project manager for two separate office buildings. The outside architectural firm, Albert Kahn, oversaw the Kansas City plant office's modernization, and GM's Argonaut (realty) Division worked on the modernization of the St. Louis office. Simultaneously observing the competing engineering abilities of these two respected architectural engineering firms was a great experience. Albert Kahn was the best architectural engineering firm in the country during the 1960s.

This truth was confirmed when I was working with the CCO Director Bill DeGrow. Accused by corporate management of using only Detroit architectural engineering firms, Mr. DeGrow chose me to question two of the world's largest firms located in New York. I reviewed their latest drawings of a manufacturing plant and asked pointed questions to expose their technical expertise. DeGrow later said, "They just don't compare to the Detroit firms at all." I am still proud of the fact that DeGrow relied on my expertise to evaluate these firms.

I believe it was in December of 1967 when Bill DeGrow came to my office, stood in the doorway and said, "You are going to the Corporate Plant Engineering Group."

"Will I be promoted to level eight from level seven?" I asked.

"No! You are being double promoted to an unclassified position," he said with a smile on his face. This wasn't standard procedure, but they made it work.

As I moved from the Chevrolet Division to the corporate level, I brought with me the knowledge and experience I had gained from studies I had conducted for Chevrolet.

In the early 1960s, I worked with a weatherman and an NYU professor to conduct the first pollution plume studies. Chevrolet foundry loading dockworkers were collapsing due to carbon monoxide exposure from the door heater units but by studying the path of the exhaust plume, we were able to solve this serious health problem. This experience prepared me to offer feasible options for GM plants where the direction of the exhaust plume was of concern.

In 1966, the Los Angeles "Rule 66" passed for hydrocarbon control (HC). Bob Phillips and I used copper tubes and buckets of snow to conduct the very first tests on the spray booth emissions from the Chevrolet Flint truck plant. We recorded high HC readings every time the spray guns went off. The health risk was finally coming to light.

Something I will never forget is seeing the worker on each side of the car body being covered with paint from the painter on the opposite side. It was within the decade that robots were applying the paint, but the issue of spray booth emissions became one of GM's biggest headaches and the very concern that the Paint Plan was supposed to address. A complete explanation of this issue is laid out in the chapters to follow.

## Corporate Plant Engineering (1967–1972)

When I was promoted to Corporate Plant Engineering there were only six of us on the staff. It was our responsibility to deal with the engineering problems that might affect all GM plants. Examples would include compliance with new electrical standards, proposed government regulations or operating codes. We were charged with the responsibility of reducing corporate energy usage and initiating noise and air pollution policies for all plants. We also served all the divisions as the lobbyists and liaisons between the corporation and the plants, and the regulatory agencies. From 1967 to 1971, our boss was the forward-thinking Howard Baldwin. Before I moved to the corporate division, I had been enlisted by Baldwin to work on a special project that involved searching for cleaner ways to burn coal. I had heard his predictions for a future energy crisis, future waste disposal issues and the need for clean energy. No one else at GM was anticipating these issues, at least not out loud.

Then in 1971, Italian born Gabriel Tiberio was promoted from staff engineer to the director of Plant Engineering. Tiberio was highly regarded by many. He was the one who sold the programs and got us the resources. Together, Gabe Tiberio and I implemented GM's corporate noise abatement programs as well as corporate programs to address issues of air pollution, water pollution, and water usage.

Though we generally made a good team, once he was in charge Tiberio made it clear that his opinion was not to be debated. I believe the maxim to be true that the most intelligent thing two human beings can do is debate an issue and learn from each other. So when I said, "Did you consider this?" or "What if this happened?" all hell would break out. The meetings he conducted should have included time to stimulate ideas, but different concepts were rejected or immediately opposed. Tiberio was the first of my bosses to squash critique. Some of the other executives in these meetings would stop by my office later to compliment me on my endurance and determination. I may have been trouble for him, but he always gave me more to do.

## Environmental Program Manager— The Environmental Activity Staff

In 1972, with the rise of the EPA, the Corporate Plant Engineering Department morphed into the Environmental Activity Staff. It was the same group of engineers still working under Gabe Tiberio but regulatory compliance became our full time focus. In time, all environmental programs were placed under my management. This GM department was respected in the industry. Our section's supervisors chaired most of the Motor Vehicle Manufacturer's Association committees and represented the Association's members in Washington DC. It was a great team of engineers; they were extremely busy but happy to be at the cutting edge of environmental initiatives. The early seventies were exciting times, with new precedents being set every week.

Through GM's environmental consultant, Dr. David Gates, I was introduced to Dr. John Cairns, a water microbiologist on the faculty of Virginia Tech. He had developed bioassay techniques to determine sources of pollution. The studies would gather data on the diversity of biological life (insects, fish, plants, etc.) for the purpose of determining if there was any unhealthy reduction under the influences of pollution. I had Dr. Cairns and his associate, Dr. Dickerson, conduct background bioassay studies at four different types of manufacturing plants. My establishment of these baseline bioassay studies paid handsome dividends for GM. The reports were sometimes shelved for years but when pollution concerns arose at these plants in later years, comparisons with the baseline studies helped to pinpoint the pollution source, or in two cases, prove that the plant was not the source of the problem as regulators had been quick to assume. In one case, I had received an anxious call from David Skivens of Fisher Body concerning the Flint Coldwater plant, but to his relief, comparison of a recent study to the baseline numbers revealed that there was no measurable adverse impact despite a known issue with wastewater treatment. In the second case, initial findings of lead in the soil around an Indiana Battery plant were determined to be naturally occurring in an unusual strip of soil that extended all the way to Chicago. The threat of heavy fines was rescinded as the studies gave proof that the plant was not the source of the problem. It was not long after this that environmental impact studies, including bioassay studies, became an EPA agency requirement for new building sites. I am especially proud of these first studies.

In 1973, the oil embargo was enacted and GM had to close all its assembly plants for some weeks because of poor sales. There was a ripple effect throughout the corporation due to the financial strain. One result of the oil embargo was that Gabe Tiberio was transferred to form an energy department to address corporate energy usage, and another effect was a major shift in attitudes toward regulations. The government came to be considered the enemy of the industry, with regulations that increased production costs at a time when corporate profits were down. The Divisional Central Offices adopted the practice of impressing management by resisting further environmental regulations. Some of my best people stopped aggressively solving pollution problems and began resisting them at all costs. Resistance pushed beyond logic. The corporate climate encouraged resistance even when addressing the problem could reduce costs or make a profit for the corporation.

My thinking was different from many of those within GM and increasingly we were on a divergent path. Many engineers could not distinguish between management's desires to avoid unnecessary expenditures and the long-term interests of the corporation. From my early years in small businesses, I had been trained to consider the effect of a decision on long-range profitability. In reading my opening chapter, some have declared that I was not a "team player." This very charge was indeed leveled against me during the period of the Paint Plan and resulted in a summons from Howard Kehrl, the vice chairman at the time. No one would have said this about my time at Gabriel Steel or my early years at GM. Did I change? I think not. The game around me changed and I was unwilling to play by the new rules because they didn't make sense to me.

Subsequent to my two and a half hour meeting with Kehrl, his chief of staff called to tell me how pleased the vice chairman was with my report. I was, like numerous decent and smart people within General Motors, struggling to survive in a tough economic environment.

This then is the path that shaped my perspective and trained my mind. I have no regrets and no sense of ill will toward those who failed to see the bigger picture. After spending 35 years in the employment of GM, I know that some of the problems revealed in this book still remain unaddressed, but I haven't lost hope. At the conclusion of the book I have outlined several steps that the corporation could take to move forward and increase its profitability.

In his 86th year, Michelangelo wrote "Ancora Imparo"—"I'm still learning." My work experience has taught me that the best men do not become so because of money or social standing, but because of their willingness to recognize problems, fix them and learn from them. My success within General Motors is attributed not to my salary or position in the company, but to my open-mindedness and willingness to share ideas and improve myself. Leading a fulfilling life means continually educating yourself and being open to being educated by others. It means having the courage to think differently. The founder of GM thought differently than many around him and I believe GM can yet recapture some of the thinking that made it great.

# 3

# An Overview of GM's History

## Formation and Growth

The founder of General Motors, William R. Durant, or "Billy" as he was known, was a successful salesman from the world of the horse drawn carriage. The company he was with was the largest maker of carriages in the U.S.[3] While Henry Ford was creating a business model that specialized in one design, the Model T, Durant envisioned creating different products at different price levels for different populations of customers. He had seen this strategy work in the carriage market. To achieve his vision, Durant pulled together 13 start-up automobile companies including Buick and Cadillac and 10 parts suppliers, and in 1908 he created the General Motors Corporation.

Also in 1908 two of the seven Fisher brothers, who had been building horse drawn buggies in Indiana, moved to Detroit to gain a foothold in the infant automobile business and soon the family founded the Fisher Body Corporation. Their specialty was not the engine, but, as the name implies, the body of the vehicle. Vehicles before WWI were open-topped and square-sided, with the only curved surfaces being the fenders. A box on wheels, painted black. To increase sales, the passengers had to be protected from the weather. In 1910 only 10% of vehicles had enclosed bodies. By 1930 the majority of vehicles were all-weather.

Though initially made out of wood, the Fisher bodies were successfully designed to withstand the additional shaking of the mechanical engine. Out of their Fisher Body assembly plant, designed by Detroit architect Albert Kahn, they provided bodies to most of the major auto manufacturers of the time. In 1926 the two corporations joined forces and, for better or for worse, the Fisher Body Division became an inseparable, although distinct, part of General Motors.

After a period of leaving GM and then returning, Durant brought in the Chevrolet Company as an additional division for the economy end of the market and he brought in key personnel in the form of Charles Kettering and Alfred Sloan. Kettering was an innovator, who invented the self-starter (as opposed to an external crank starter) and who discovered the increased engine power that can come from use of leaded gasoline (without realizing the health hazard that it posed). Kettering is also credited with taking Duco nitrocellulose paints (lacquers) patented by DuPont and adapting them for a quick drying application to automobile bodies. His advances opened up the world of colored paint to the auto buyer.

Alfred Sloan was an organizational guru whose structuring of the corporation is still studied in business schools today. Together, the philanthropy of these two gentlemen funded one of the primary research hospitals in the fight against cancer—Sloan Kettering Memorial Hospital in New York.

Under the guidance of Sloan, the combination of talent and market saturation that Durant had assembled was wildly successful. Alfred P. Sloan, Jr. was president of GM from 1923 to 1937 and then chairman of the board from 1937 until 1956. He and William Knudsen, who followed Sloan as president, are credited with creating a "stair step" of GM offerings for every level of buyer and the incentive to trade up to new models in order to take advantage of annual model changes that introduced new features and product options.

Following World War I, two historical events challenged the way the automotive business was conducted. First, the 1929 stock market crash and the ensuing Great Depression taught valuable lessons. It forced management to curtail production, reduce manpower, and analyze every expenditure for its cost benefit while operating daily on the brink of bankruptcy—hard lessons on how to operate in the worst of times.

Second, World War II forced industry adjustments. Management had to cancel some automobile production in order to produce military equipment. Together the U.S. automobile manufacturers and the thousands of Detroit tool and die shops that they supported formed what was known as "the arsenal of democracy." Factories operated 24 hours a day, 7 days a week churning out tanks, planes, and military trucks with limited time for the maintenance of equipment. The company was pushed to work within government specifications and quality control standards. Production demands were high and with so many of the men overseas, GM brought women into the workforce for the first time. The war years were trying for all, but they taught the industry how to adjust in turbulent times and still make a decent profit. An effort could have been made to retain these lessons in the corporate memory, but as too often happens, most of these lessons were lost over time.

But in the short-term, coming out of the war, General Motors management, as established by Sloan, was at its very best, and the company's number one standing in the world, by every standard, was proof of its superiority. Back in 1923 when Sloan took the helm, Ford held 60% of the auto market to GM's 12%, but under his leadership, GM became the industry leader. Alfred Sloan continued to operate the company as an umbrella over autonomous divisions, each charged to develop cars for a distinct market niche. Only financial and policy controls came from a central office. Each division had its own teams for design, engineering, and marketing and then utilized Fisher Body to produce the body of their vehicle. The historic GM divisions were Chevrolet, Oldsmobile, Pontiac, Buick, GMC, and Cadillac, each with its own personality and loyal employee and customer base. Every division had a distinct style that could easily be recognized, as in "that's a Cadillac" or "there goes a Buick." The tail fin rear fenders, the Chevrolet front grille, the false exhaust ports on the Buick's engine hood, etc. made every division's vehicle distinctly different and easily recognizable. Under this highly acclaimed business plan GM accomplished Sloan's oft repeated goal of "a car for every purse and purpose."

In the years to follow, General Motors set the benchmark for styling with the cartoon-inspiring tailfins, the Pontiac wide-track, the four-door hardtop, wraparound windows, and the Corvette dream car. GM's technical innovations drove the market with independent front wheel suspension, a one-piece steel roof, the Rocket V8, automatic transmission, and the first power steering, all demonstrating the company's leadership role. By this time, GM sales had outstripped those of Ford with GM now holding over 60% of the market, and there was talk of breaking it into two companies, similar to the split of Standard Oil into smaller companies a few years earlier.

By the late 1950s, GM was the most profitable company in the world. In the 1960s, GM was riding high, expanding its manufacturing and assembly plants to accommodate increasing demand and the standardization of accessories like radios, air-conditioning, and automatic transmissions. In 1966 in Lordstown, Ohio, GM built its first new assembly plant in years. In addition, approval was granted for a new nodular-iron casing foundry in Saginaw, Michigan, and a large Fisher Body stamping plant in Kalamazoo, Michigan. There was talk of doubling the number of shares to reward investors. But the entire industry was on the brink of major change. The government would soon be implementing regulations that required vehicles to be safer for the environment and safer for the passengers. In addition, a new business threat appeared with the arrival of Japanese imports with their claims of higher vehicle quality and greater fuel efficiency—claims that proved to be true.

But while external pressures were amassing just over the horizon, there was a sense of complacency that set in after a decade of notable success. As in any organization or group, there is great unity of purpose when the focus is on an external threat or a clear common goal, but when the goal is achieved and the enemy is vanquished, almost inevitably, internal strife rears its ugly head. The divisions that had once focused on total corporate growth had hardened over time into independent fiefdoms sometimes lapsing into either questionable ethics or cronyism. I had the opportunity to personally observe the manifestations of these internal struggles and also the attempts that management made to address the issues through a process of centralization. Not surprisingly, the division with the reputation for being the most prideful and independent was the Fisher Body Division.

## Fisher Body: Internal Troubles Begin

In the simplest terms, the automobile business has one objective: to produce vehicles that the buying public considers of greater value than those of the competition and to do so at a profit. A car buyer perceives value in a vehicle's overall appearance, usefulness, and its quality of workmanship. The reputation of the company and the vehicle's name recognition are also factors. The one GM division responsible for most of these value characteristics was Fisher Body.

All five car divisions had separate rooms in the GM Tech Center design studio building. The car division determined the overall size, shape, weight, and chassis for a new model, and the studio prepared beautiful renderings and clay models of the vehicle for approval. Once approved, the renderings would be sent to Fisher Body to design the actual body parts for manufacturing.

Before the modern era of 3D computer images, Fisher's model shop would produce a full scale wood model of the vehicle for the designers to check their parts against the drawings. Fisher Body model makers and designers were some of the best in the business. Fisher's tool and die room could construct almost any part. The actual cutting of the steel punches and dies was a difficult machining job. The die cutting machine had a tracer arm, which rolled over the wooden model, directing the tool which was cutting the metal die. It was an effective system given the available technology of the times.

In 1967, Fisher Body had plants in 27 cities making the large body panels, trim and miscellaneous small parts. The large stamping plants with sheet-metal presses made hoods, fenders, roofs, trunk lids, floor pans, and door panels. In large volumes, the parts would be shipped by train or truck to one of the various Fisher Body assembly plants located adjacent to its companion car division's final assembly plant. Note that the cost of shipping is a significant cost consideration and in some cases more expensive than the manufacturing.

The home or main offices of the five car divisions were located in Michigan. Cadillac was in Detroit, Pontiac in Pontiac, Buick and Chevrolet in Flint, and Oldsmobile in Lansing. Their Fisher Body assembly plants were all located blocks away from the car division's plant with multi-deck haul-away trucks (carrying 12 to 18 bodies each) traveling between the plants most of the day.

Fisher Body assembly plants, receiving the panels from the press plants, would spot-weld them together, wash the body in white, prime and color coat it, and add interior trim before sending it to the car assembly line through a hole in the wall dividing the two divisions. They belonged to the same corporation but treated each other as competitors.

Every vehicle was made to a dealer or company order with color, trim, and options all pre-specified. The car division production line operated at a rate of 60 to 70 cars per hour so the bodies coming from Fisher Body had to be scheduled at exactly the right moment to be dropped on the correct chassis (the main frame of the car with the wheels and engine already in place). Cooperation between the two divisions was essential for an efficient operation and it was a stress point at times.

The first damaging crack in the previously successful GM structure came between Fisher Body and the car divisions. The Fisher Body Division made their profits on the backs of the car divisions. First, they were paid up front for the design and body production without regard for actual sales. Then once the initial design had been submitted, the cost of implementing any changes was charged to the car division. Fisher Body was guaranteed a 33% markup for all the additional time, labor and expenses they incurred when making an alteration to the original design. Designers, production engineers, and often upper management were known to change the original design to meet their personal preferences, and Fisher Body counted on this extra profit. There were constant disputes between Fisher Body and the car divisions related to this high profit mark up and concerns about the quality of the vehicle bodies. The vehicles went out with the moniker of a particular division, but the leadership of that division had no direct control over the body they received from Fisher.

Fisher Body had control of so much of the process of bringing a model into existence that it became a car company within the company. It had its own ways of doing things, many of which irritated the car divisions. At times, Fisher limited desirable trim options or conversely allowed the production of body styles that were not expected to sell much volume. There was constant internal friction. The two groups were interdependent but never seemed to work in harmony, rather protecting their turf for the purpose of self-interest.

In an effort to see if they could work in a better partnership, the new 1966 Lordstown Chevrolet assembly plant was built with both under one roof, with a joint plant maintenance agreement to save each division from having its own clean-up crews. But the experiment failed. Fisher Body and Chevrolet fought like competing companies, even refusing to pick up each other's trash.

The sense of entitlement within Fisher Body was infamous and spilled over into its relationships with suppliers. To many, it was apparent that corruption had taken root in the Fisher Body division. Back in 1957, when I began my GM career at the Chevrolet Engineering Center, I was working across the road from the front door of the Fisher Body central offices. Within days, I noticed large numbers of Fisher employees being taken to lunch by outsiders. I found it was well known that key positioned employees were receiving gifts delivered to their homes, tickets to sporting events, invitations to golf outings, and vacations around the country from suppliers and contractors.

I also had direct knowledge of the relationships between GM engineers and the paint and solvent suppliers. Like many in the company, I felt that this area needed to be investigated. I recall a plant engineer in Flint, Michigan, who was exposed by his wife in their divorce proceedings for receiving cash each week at his home from a paint supplier. She wanted her share.

In another example, Fisher Body's paint manager retired from GM and proudly went on to accept a six figure salary, a company car, and a golf club membership in Atlanta, Georgia, from a paint supplier who hoped to profit from the man's connections with the paint research center at the nearby GM Doraville plant.

Overall, Fisher Body's relationship with vendors and contractors was infamous; yet was never stopped until the early 1990s when a new Vice President José Ignacio "Inaki" López de Arriortúa established firm ethical rules of conduct for the taking of gifts and free trips from suppliers, and he backed up these rules with enforceable consequences. In the '70s and '80s other methods of dealing with Fisher Body were put forth.

## Centralization and Reorganization

In an effort to address these problems, Fisher Body and the car divisions at GM were reorganized a number of times. Initially it was a matter of centralizing for efficiency. In overall appearance, all assembly plants look alike and operate using similar equipment. To save money and people, why not centralize some of the engineering services? So in the early 1960s, a small GM assembly division began operating in the GM building in Detroit serving Buick, Oldsmobile, and Pontiac assembly plants. The services provided to the plants related to the physical building services such as electrical, mechanical, and powerhouse systems. They dealt with specialty issues, not day-to-day operations. During this time, I was working for Corporate Plant Engineering but our office was handling regulatory compliance issues, not plant engineering issues.

For daily routine engineering problems, every assembly plant had a plant engineering department, but the GM assembly office also collected data related to energy usage, compressed air demands, and powerhouse data for a comparison of the assembly plants as a way to improve efficiencies. As I will detail later, GM underutilized this data and in many critical areas failed to even collect it.

Recognizing the benefits of this first step toward centralizing certain services, management made a greater step toward centralization in 1969 by appointing Joseph E. Godfrey as the official vice president and general manager of the fledgling General Motors Assembly Division with the responsibility for plants in eighteen cities.

Over the next four years, from 1969 to 1972, the assembly plants were transferred from the car divisions to the GM assembly division in phases. Godfrey began to take more and more control implementing cost control measures, which as former GM executive, John DeLorean states were against the better judgment of the divisions because they resulted in lower quality and hence higher warranty expenses. The quality goals of the car divisions were undermined by the cost cutting goals coming out of GMAD.[4] The engineers in the plants were never consulted by the engineers working directly for GMAD in a central office. Decision-making became too far removed from the actual manufacturing process.

During these years, the lack of cooperation between the main divisions persisted, so in 1973 Godfrey had management appoint him vice president and group executive of both Fisher Body and the GM Assembly Division, bringing Fisher Body under his charge as well. This was the second reorganization. Robert W. Decker was appointed as vice president of Fisher Body Division and Charles Katko as vice president of GM Assembly Division, but both now reported to Joseph Godfrey.

## Changing Times for the Industry

In the 1970s, globalization of the entire industry began to occur. Every automobile manufacturer in the world recognized the North American auto market as the largest and most lucrative. Through this time period the U.S. demand for automobiles skyrocketed and seemed insatiable, on its way from annual vehicle sales of over a million to sales of 17 million. An assembly plant in the U.S. operating at full capacity had (and still has) the potential to generate hundreds of millions, even billions in profit for the corporation. Companies outside of the U.S. wanted a piece of the pie, whether that came from the manufacturing of parts and components or entire vehicles. The lower costs of overseas labor and raw materials began to be a critical factor in a competitive market, but there was also a drive and focus in these younger companies that had been lost at GM.

With the additional financial stress brought on by the oil embargo of 1973 and the resultant temporary plant closings, a program was put in place to eliminate many of GM's small parts manufacturing plants. Fisher Body finally conducted detailed cost analysis of each part made in their trim and small parts plants. They discovered one or two parts generated great savings (made money) and a handful of parts broke even, but most were made at a loss and could be purchased from outside the company at a much lower cost from overseas or from non-union U.S. businesses. GM retained a few foundries, large press plants, and a handful of engine and transmission operations, but largely the focus of the corporation narrowed to the design, assembly, and sales of vehicles.

This change in business focus was critical, yet GM's underlying culture was still the same, and during these years the maxim proved true that "pride goeth before a fall." The unresolved problems in GM's culture that surfaced in the 1970s can be traced back to its once proud status as the biggest and most profitable company in the world. Even the GM website, in its presentation of corporate history, states in regard to the pressure to produce cars that compete with the Japanese models, "GM rushed to develop smaller vehicles as well, but the company had been too large and too successful for too long to change direction easily, and GM's undisputed dominance of the U.S. market began to erode."[5] In the July 6, 2009, edition of *Fortune* magazine (p.82) Akio Toyoda, the new president of Toyota Motors was asked if he was worried "about the company growing so large and bureaucratic that it becomes infected with 'big company' disease." He replied, "I believe we will be able to figure out ways to avoid being infected." He was reflecting on the demise of GM and the Chrysler Corporation.

"Big Company Disease" was certainly a problem that plagued GM. Management and the board of directors may have thought that combining the Fisher Body division and the car division under one leader would put an end to all abuses at Fisher Body. Sadly, management gave all central office functions in GMAD to the Fisher Body people, and the principle of self-interest over and against the corporation's best interest continued unchecked, with its escalating costs hidden in the voluminous corporate balance sheet.

Former Fisher Body management controlled most plant operations from central headquarters at the tech center. The GMAD organizational structure failed to produce the expected improvement in cooperation between the various groups. There was potential in the centralization process to move GM ahead, allowing it to maintain its position as industry leader, but it didn't happen.

Imagine, the newly formed GMAD could have built a talented staff of brilliant engineers from MIT, Berkeley, and other universities in the forefront of technological development. It could have hired the best process engineers from the tool and manufacturing companies serving GM and have had the leading robotic and computer engineers on staff. It could have consulted with the most experienced and intelligent of the GM plant engineers. I worked with several of these engineers in the Chevrolet division. Men like Jack Cragen had impressive credentials and years of plant experience upon which to draw.

GMAD's central office could have had analysis engineers collecting operating data (labor, material, energy, maintenance costs) from every plant for comparison. It could have been taking the best, improving each operation, and measuring performance in a "continuous manufacturing improvement program." It could have been applying new technologies based on engineering performance assumptions, testing those assumptions, and making adjustments based on observed performance. But none of this happened. Billions of capital was invested in new construction and paint shop upgrades without the most basic of engineering principles being applied. The final "new" paint shop built under the Paint Plan was replaced in 2013 because it was an efficiency failure from day one. As the final shop of 31 it should have been the culmination of all that was learned in the upgrade process. Instead it has been one of the final standing monuments to poor planning, poor design, and poor engineering.

## More Reorganization

In 1984, under Roger Smith, another reorganization took place. Under pressure from outside sources (not the board of directors), CEO Roger Smith and President James McDonald decided to do something big and eliminate Fisher Body and the Assembly Division, or as reported in the newspapers, "dynamite them out of existence." This drastic restructuring was an attempt to internally combat the "big company disease" and also address the conflicts between the divisions. So in 1984, the long-standing name badge divisions were grouped into the Chevrolet, Pontiac, and GM of Canada Group (or CPC), and the Buick, Oldsmobile, and Cadillac Group (or BOC). To further disrupt any prior power structure, Robert Stempel, the former manager of Chevrolet, was appointed to head up, not the CPC, but the BOC and Lloyd Reuss was pulled from managing Buick to head the CPC.

The combining of the assembly plants was a massive restructuring program. There were tremendous risks involved and the badge loyalists in the car divisions were predictably fearful of losing control. The logistics of dividing the pool of automotive engineers in two proved to be an onerous task. Anyone from GM management at that time will affirm what DeLorean revealed to author Patrick Wright—it was a time of tremendous angst and it resulted in more inefficiency, not less.

Despite Smith's best intentions, the inner circle of team players, most from the Fisher Body operating group, retained control over auto assembly plant operations and worked together out of the CPC division. Fisher Body's D.D. Campbell, despite his poor track record, was fully in charge of CPC's operating groups.

And while R.G. LeFauve was titled as director over the various engineering groups in BOC, the BOC division only had a very small assembly plant operation group. In matters of plant construction, operation, and implementation of the Paint Plan, they too still reported to old Fisher Body staff operating within the CPC division. LeFauve told me directly that his orders regarding big programs, like the Paint Plan, came from D.D. Campbell in CPC; so the structure was reinforced when within four years, LeFauve had been moved to head Saturn. The 1988 management flow charts show that all "Operations" were led by Campbell out of the CPC group. The organizational charts kept changing, but not the people causing the most harm.

The combining of the assembly plants and the changes in Fisher Body's division started in 1967 and didn't stop until 1992, with a number of major modifications in between. None of these attempts by management to correct problems ever got to the real source: a core group of people who had come from Fisher Body were still in control of these divisions and the culture in which they operated. In fact, the 1984 CPC/BOC reorganization made things worse for the corporation. I believe "outsiders" like Bob Eaton and Alex Mair, and other good men could have led the car divisions in a better direction, but they were never given the opportunity.

It should be noted that in that same year, 1984, Roger Smith led the charge for GM to buy Electronic Data Systems (EDS), owned and operated by Ross Perot, and the next year, Hughes Aircraft. As documented by several industry observers, the purchase of EDS was intended to give GM an edge in expanding computer usage in all areas of the business, and the purchase of Hughes was justified as business diversification, but the integration of EDS in particular only increased employee disgruntlement and served as yet another distraction from GM's core business.

No one would accuse Roger Smith of sitting idle during the years of the Paint Plan. Yet many of the projects that he initiated, while sounding impressive on paper and in the press, distracted the board, drained the corporate coffers, and had little if any connection to the customer—GM's primary source of income. The pride remained but the corporate standing of the 1950s had largely evaporated by the end of the Smith years. This then forms the backdrop for all that went wrong with the Paint Plan.

# 4

# GM's Leadership Woes

A review of GM organizational charts will show that there were numerous reorganizations through the years, but always and to this day, the highest governing body is the board of directors, composed primarily of outside "experts," along with the chairman of the board/chief executive officer (COB/CEO), and the president of the corporation/chief operating officer (COO). Under the board and these two leaders are several executive vice presidents (ranging in number from two to six), then a level of nine "Vice Presidents and Group Executives," and reporting to the group executives are all the other vice presidents. The problems that I observed and the resultant financial losses came predominately out of the poor practices of the assembly division combined with the prideful attitude of the Fisher Body leaders. They were spearheaded by a few key players, but corporate cultures begin at the top.

Although it was evident at all levels, the problem of pride had its origins in the attitude of the board and chief executives. A sense of greatness: "we are the very best," "who can tell us how to run the business" reinforced this culture of superiority within management. It engendered the notion that any smaller company cannot possibly give good advice. In different decades, both John DeLorean and Jack Welch spoke out on how this feeling of superiority caused the leadership not to take outside advice. Jack Welch, former CEO of General Electric, was quoted as saying in a 2009 Detroit radio address that "GM's problem is they won't listen to advice by those outside the corporation." More stinging words came from former Vice President John Z. DeLorean, who observed in the '70s: "A fault that GM has had for a long time is its feeling that since it sells more cars and trucks than anyone else in the world and makes more money than any other automotive company, the GM way is the only way."[6] I personally observed several examples of this at both the board and upper management level. Consider first the story of the corporate ecologist Dr. David Gates.

## Expert Advice Ignored

It was in the early '70s, when the members of the Environmental Activities Staff were eagerly and happily working on the cutting edge of environmental issues that I met a wonderful scientist, world traveler, and dear friend, Dr. David M. Gates, a University of Michigan professor newly hired by the GM Board to act as their consulting ecologist. It was my good fortune to be assigned as his guide on most of his monthly visits to review GM's research projects, atmospheric studies and environmental programs. He was a fascinating man. Dr. Gates was on the Leakey Foundation in Africa, studied the environment at both poles of the earth, knew men who flew around the world on one tank of gas, and crossed the English Channel on a propeller blade-driven bicycle. He has taken the Orient Express across Europe and has written and published 11 books. Dr. Gates expanded my view of the world and made my years at the Environmental Activity Staff (EAS) more exciting. We remain close friends to this day. But the lack of interest in Dr. Gates by most GM management was appalling. He and I would spend his assigned day together, but the upper management who had hired him to give opinions would only give him a brief hearing over lunch. His first GM plant visit was to the site of the Oklahoma assembly plant. He made a recommendation to Joe Godfrey's GMAD people, and the GM real estate Argonaut staff that accompanied him, that the external environment should be left natural looking—natural plants and flowers. Shortly after his visit all the plants were removed.

Dr. Gates was very knowledgeable about energy futures and was a climate-warming expert, yet in the 16 years that his services were retained (1972–1988) he was never asked to make a report to the Environmental Activity Staff or the Public Policy Committee on these important public issues. In fact, to my knowledge he was never given a specific assignment or asked to give his opinion on any important matters. Initially, he wrote one or two reports for a Public Policy committee of the board, but that was stopped early on. In truth, the only way that GM utilized this knowledgeable man was to cite his name as a means of fending off a stockholder proposition for GM to be subjected to external environmental impact studies. The underutilization of Dr. Gates was a clear example of the manner in which the views of outsiders were ignored and unwanted. It was a symptomatic show of GM arrogance and maybe even ignorance.

## The Unqualified Board

Perhaps it is pride in a successful history that led the new GM Board to a place where all members were chosen for their knowledge of financial matters, but not a member among them had engineering training or manufacturing experience. From the February 16, 2010, issue of *The Economist*: "Rather than making changes that actually improve corporate governance, boards are made more formally independent, without actually increasing their ability to control management, typically done by hiring directors who, although they may have no business ties to the company, are socially close to its top brass."[7] This practice of hiring financial gurus of high social standing was a departure from the philosophy of respected president, Alfred Sloan, who was quoted as saying, "I happen to be one from the *old school who thinks that knowledge of the business is essential to a successful organization.*"[8]

Honda, the best managed of all the car companies, according to a 1981–82 study initiated by Iacocca at Chrysler, has a predominance of product and manufacturing experts on its board, many of them former company executives. Honda's board of directors and senior corporate management stand out for their intimate knowledge and experience with product programs and manufacturing systems. In contrast, GM's current board is composed of traders from Wall Street's private investment firms, and executives from a railroad, from communication companies like AT&T, and from information companies like Ceridian Corp. These persons are financial experts; not one has expertise in manufacturing or automobiles. I observed in meetings, conversations, and in the resulting decisions, that both the board and top management rarely understood the complexity of automotive assembly and paint operations, or the cost of plant construction. Boards of directors fail a company when the majority of members lack real knowledge of the company's production and operating systems, and the few members that do have knowledge, owe their allegiance to the president or chairman of the board (COB) and therefore aren't free to speak critically. Shouldn't boards be used to debate and decide the most important issues facing the company?

So to the company's detriment, GM's board members were not capable of analyzing the outrageous promises of the 1978 Paint Plan. This plan involved the massive conversion of all of the GM paint systems within 10 years, with as yet undeveloped paints and untested paint systems and application equipment. The board's willingness to approve the Paint Plan without a serious debate about its schedule of plant shutdowns, and without regard to product programs and possible economic fluctuation, clearly indicates that the board was unqualified for corporate leadership. Their lack of engineering and manufacturing expertise resulted in the approval or unchecked allowance of a host of profit draining programs that I will consider, with the pinnacle of folly being the 1978 Paint Plan.

This level of ignorance also caused the board to be treated with disrespect within the corporation. Consider the following event, which I observed first hand. It involved giving information to the board of directors. In 1979, the new director of plant environmental programs for the EAS, Dave Edens, hired from outside the corporation, wrote an introduction to his staff annual report to the board. For the first time in eight years, the board did not receive a simple list of all the plants that had compliance programs already underway—which the board could do nothing about—but also laid out the complete picture of how the environmental movement had affected GM's plant operations and would continue to do so far into the future. It presented an overview of the actual costs of materials and pollution controls resulting from government regulations. The information was clearly presented. Any intelligent person would enjoy reading its contents and it would seem to be of importance to the directors. But a fellow from the treasurer's office said to me, "Nice writing, but we write to the board as if they were third graders." The director had put his whole heart into the introduction and decided never to write another, giving the task instead to his managers and me.

When I took over the task of writing this report to the board of directors, the quality of my report introduction was never at the level of Eden's, but I too felt compelled to bring to the board's attention serious environmental problems that GM needed to address. Every year since the EPA was formed, an annual report to the board from the environmental staff would list all the plants that had approved EPA compliance plans. Instead, I wanted to make them aware of community complaints and the three plants that had PCB contamination problems. These three situations had resulted in multiple lawsuits, the resolution of which was estimated to cost GM tens of millions in unplanned funds. Specifically, the Massena, New York, plant pollution cleanup had already cost the corporation over $100 million with no end in sight. *Is that the sort of subject the board should be aware of?* I thought so. But Dave Potter, my executive vice president, resisted and ordered me to have each plant group vice president sign off or drop the idea of highlighting their plants. Instead each vice president easily approved my proposal, thinking everyone in the world knew about his plant problems: the plant workers, the citizens suing GM, the state agencies, the EPA—everyone except the board and my vice president. But apparently keeping vital information from a board of directors is common in the U.S.

## The Uninformed Board

The August 4, 2009, issue of *Newsweek* stated that it believed the Chevron Co. board and their stockholders knew very little about Chevron being sued in the past years for polluting the Amazon rainforest.[9] This doesn't surprise me because a friend who was a Detroit Edison board member was misled by management about a serious pollution problem that the utility giant had. Management had unreasonably pressured a group of Air Pollution Control commissioners in order to win the approval of their pollution reduction plan.

My friend spoke to me right after his board meeting, saying, "Edison has agreed to reduce their permitted sulfur coal content from 2.3% to 2.1% sulfur (S), even though there is no proof of Edison being the air quality problem."

A simple fact check shows that the Edison Monroe power plant burns a 100-railcar trainload of coal a day, whereas GM's fifty-five power plants all over the U.S. burn less than one 100-car trainload a year. Computer models clearly indicated that Edison emissions were exceeding the Air Quality Standard for sulfur dioxide ($SO_2$) at several locations. The board member was operating in the dark.

Without hesitating, I responded: "Edison will be at 1% sulfur coal limit within months." I knew this fact because on GM's behalf I had attended and made presentations to the Air Pollution Control Commission throughout the '70s and later had been appointed by Michigan's Governor Milliken to serve on the Commission. But I should not have known more than a director on the Edison board. What he didn't know is that the Air Pollution Control Commission meeting in regard to the Edison plant broke many public meeting laws by nearly eliminating the public comment period. Three of the commissioners, pressured by Edison to vote a certain way, feared the public's wrath and ran out the back door as soon as the vote was taken.

Three months later, Edison was ordered by the same commissioners to burn coal with 1% sulfur content or less. The deception of the board members included statements by management that 1% sulfur coal would cost Edison well over $500 million every five years. The state agencies and public were told this same figure. What they were *not* told was that Edison never took bids for low sulfur coal before stating the $500 million-plus cost. Months later, I was told by the Edison chairman of the board that the 1% sulfur coal, now under long-term contract, was millions of dollars cheaper and of a much higher quality. It had a higher BTU, "British Thermal Unit" that measures heat production, so the cheaper coal proved to produce more heat. Instead of being an extra expense, the stricter environmental standards were going to save the company money, but their board didn't have enough knowledge to foresee this truth.

Corporate leadership needs to know early on when an issue is important. This allows them time to allocate staff and resources to find solutions ahead of negative repercussions. In the 1970s Ford Motor Company's engine fan blades flew off, resulting in the fatality of a person who had been looking over the front part of the engine.

An investigation found that the engineers knew of the problem, but to save the company money, decided the risk of occurrence was low and worth taking. Henry Ford II reportedly came out of the meeting with the investigators saying, "God, save me from the guy that wants to save me."

Billionaire and business power broker, H. Ross Perot, also thought the board was not treated with respect because of both their ignorance as well as their lack of interest in the primary business of GM, namely automobile manufacturing. A few weeks after a fall meeting of the board of directors, over a private luncheon that I had arranged at a downtown Detroit restaurant, Executive Secretary Ann Kaplan and her boss, VP and Group Executive Marina Von Neuman Whitman, reported to me Perot's attack on Roger's use of the board. Kaplan was in the boardroom with Whitman at the meeting. Roger Smith, the current chairman of the board and CEO, left the room to take a telephone call. Whenever he was not in the room, things that were spoken were written down for him to review upon his return. Kaplan, a brilliant woman, was taking notes and remembers the surprised look on Roger Smith's face as he read the notes given to him as he re-entered the boardroom. This is what he read:

> Perot said to the other board members, "The Company is sinking and the only thing Roger Smith discusses with us is the African Apartheid situation. For months, the board meetings have been nothing but talk about Apartheid! When Roger comes back, if he wants to talk about Apartheid, let Roger go into the closet to talk to himself."

Clearly, the feelings of mutual respect that had existed between Smith and Perot at the time of the 1984 GM purchase of Perot's company, EDS, had worn thin. Not long after this, Perot was offered $700 million to surrender his stock and leave the board. I had to wonder if Ross Perot was the only director who saw a problem in Roger Smith's leadership. Why did no other director support Perot and refocus the board's attention on automobile manufacturing?

In each corporation and organization, decisions must be made regarding the level of involvement the board of directors will take in operations. It is inefficient and demoralizing for a board to micromanage, but when billions of dollars are being wasted and the company's public reputation is at risk, certainly the board should be aware of the issues and give corrective oversight. When they personally lack industry expertise, the board can only lead effectively if they have the humility to listen to the advice of those who are knowledgeable and if they have selected a trustworthy executive team that will always tell them the truth and give them the facts that they need to know.

The examples I have given have already demonstrated the sad truth that this was not the case at GM and it continues to haunt them today in the area of safety recalls. At a minimum, the board should give leadership by providing clear written guiding principles that are broad enough to guide every business decision and keep the company on a path of financial success. Remember Iacocca's study of Honda management? In that same Chrysler study, it was revealed that in contrast with the 12 to 25 (mostly financial) corporate goals laid out by the boards of the "Big Three" American car companies, Honda has only two written goals: 1) "Take care of the employee" and 2) "Take care of the customer." These are principles that every level of a corporation can use to guide their planning and decision-making. As I look back, it is telling to realize that nowhere to be found in the principles of this consumer-supported corporation were concern for the customer and certainly not for the rank and file employee.

The goals committed to writing at GM were never as immediately applicable nor as succinct. Instead there were unwritten principles that guided decision-making at GM, and one of those was the GM concept of "team play."

# 5

# Team Play, GM Style

In the best scenarios, a team player is one who sets aside self-promotion in order to contribute his or her best to the success of the entire team. This type of team play is valuable and good, but at GM team play became more about loyalty to the boss than to the corporation as a whole. Employees spent corporate time and money going to crazy lengths to honor their superiors by tending to their every want and need. High-level executives were treated like royalty. Observing, I always found lower level executives in a state of awe and reverence when in direct contact with top ranking management. It was heady stuff to deal directly with the masters controlling the wheels of one of the world's largest corporations and those at the top liked to keep it that way.

There were two major flaws in the GM management system. They were intertwined and together they contributed greatly to GM's inability to maintain its position as an industry leader. The first part of the problem was promotion based on loyalty, instead of knowledge and experience, and the compounding issue was that loyalty meant agreement at all times. Loyalty to the boss was of the highest value. Expressions of loyalty and agreement with the boss led to promotion and expressions of criticism or concern regarding any plans set forth by the boss usually led to demotion. At GM the message was loud and clear: be a team player, or else. As I will detail, this unhealthy perversion of loyalty existed at the highest levels and the corporate disease spread across the company.

## Loyalty Exalted

As John DeLorean, Maryann Keller and others have testified, GM had a skewed promotion system. Instead of matching knowledge with position, GM management believed in rewarding a "team player" for a good performance in one position with a more important position, even if the individual had very little knowledge of or experience in the new job.

Many have witnessed what happens to those questioning management and instead they make a career-saving choice to join the silent group of "company men." DeLorean observed "There were and are men in positions of power at GM who do not know the business they are running…Promotion of the 'unobvious choice' has put some of them into their position; others simply played the loyalty system game so well they just moved up the corporate ladder every few years. What developed in some cases is 'management by crony.'"[10]

In truth, GM was and is made up of many brilliant and capable engineers and managers, but the best did not always rise to the top. Some were even marked as "HI-POT" employees to indicate that they had "high potential" to become very valuable to GM, but it was my observation that some of these young employees were given supervisory positions over more experienced employees simply because of their potential, not because they were qualified for decision-making and leadership in that business area. This system might have been workable had critique and input from below been encouraged and heeded, but in a world where criticism, disagreement, and the reporting of problems were verboten, foolish decisions abounded and problems went uncorrected.

Consider the case of the director who first organized the corporate environmental activities and wrote the corporate environmental policies with great expertise. He was promoted to director of GM's Argonaut (Realty) Division and was given responsibility for the design and construction of the new plants. He was further promoted to head tool and machinery development in the manufacturing staff for being a good company man. The director's reaction to a certain letter illustrates what he understood to be a good company man. He aided the corporate ecologist's son in getting a summer job at the Willow Run transmission plant. He received a letter of thanks from the young man that included horrible stories of foremen terrorizing the employees on the production line. Rather than investigating and addressing the claims, the director quickly announced that the son was an ungrateful person for speaking ill of GM and that he would no longer pay much attention to his father, the corporate ecologist. He lost all respect for both individuals, but in hearing of this, I lost respect for him.

He was not a civil engineer nor a registered architect, nor a tool or machine designer or journeyman. Yet he had been given responsibility to supervise people in these categories. It was obvious that he had been given the position because he was a "team player."

When staff members have a supervisor that they feel confident to be the best in the business, they work harder and produce greater results. GM never seemed to realize what they were doing to staff people by asking them to report to mediocre management.

If you've had the privilege of belonging to a great team, you know the joy of seeing each member bringing out what is best in all the other members such that the sum of the group is greater than the sum of the parts. Criticism in this environment is always constructive. In contrast, GMAD engineers spent corporate hours writing critical reports about other engineers not playing their game. It was like kindergarten tattling. It was not constructive. The interests of your boss were always more important than the interest of the corporation and pleasing your boss was rewarded with promotion within the company. This is the culture that brought GM down.

Though I thought a change in the culture could be easily made, I now believe that no GM CEO to date has had the genius to reverse the downward trend to bankruptcy. Working against the top executives was the size of the company and a culture set in concrete that only a revolution could change. The problems were systemic. Games were played, keeping the "good" guy in place, promoting trusted friends or team players, without any quest for excellence. Too many key posts were filled with individuals with little, if any, experience or passion for the position. All this led to mediocrity throughout the company.

This was certainly a contributing factor to the failure of the Saturn program. As the press was told by Roger Smith, Saturn was to be "the factory of the future." With his reputation on the line, Roger Smith brazenly announced that only the youngest, brightest, all "HI-POT," engineers would be assigned to carry out this important project. They would provide new ideas, new technologies, and operate "out of the box." Not ever was it heard that the most experienced, most knowledgeable, the very best in their field of engineering will control Saturn's future. Not once!

The theory of success for Saturn was to eschew all that was old and tried, and replace it with new (read "untested"). It sounded exciting to the public, but from my perspective as an engineer, it was pure foolishness. When it came to the nuts and bolts of designing and constructing the Saturn assembly plant, was it any wonder that billions of additional funds had to be appropriated to correct the host of problems that the young "HI-POT" employees failed to anticipate due to their lack of real time experience?

Consider the case of Neil DeKoker. In the late 1960s, Neil and I met when he was making a General Motors Institute presentation to the Chevrolet Central Office Construction Group. An electrical engineer, I sat in as he presented a report to Gus Pappas giving evidence that employee work efficiency was directly proportional to office lighting. As lighting increased, efficiency increased, according to his presentation.

Gus Pappas, the electrical section manager, got to know Neil's strengths and weaknesses and at morning coffee said to me, Neil is considered a "HI-POT" employee and will be advanced too fast for his own good. Gus was an excellent forecaster and a competent engineer. One of Gus' former engineers, with over 30 years of experience, recently remarked to me, "Gus was by far the best manager I've ever worked for, the smartest electrical engineer I've ever met, and the nicest human being I ever knew."

In the early 1970s, Neil was transferred to the Environmental Activity Staff, where we met again under Gabe Tiberio, the director. After the 1973 oil embargo, Neil made a second office lighting study finding this time that one-half to two-thirds of the office lights could be turned off to save energy without a drop in worker efficiency. Gabe and I were the only ones to hear Neil present both of his light studies. It was my observations that his contradictory findings were based more on opinion than measured performance of actual workers, but Tiberio saw things differently. At the GM Tech Center and GM building many lights were eliminated and Neil was treated as Tiberio's prodigy with their careers forever enmeshed.

So, although it caught others off guard, it was no surprise to me that Tiberio named Neil to head Saturn's planning committee. His committee of 99 people from all walks of life in the company, including UAW members, presented their comprehensive plan. In an interview with Maryann Keller for her book *Rude Awakening*, Neil reflected with tears in his eyes, "'I get emotional now, just thinking about it. It was like being part of an evangelistic association—we were the Bible-wielding 'Southern Baptists' carrying our Saturn philosophy around.'"[11] There was passion and high expectation among the young appointees to the project, but no voices of experience and wisdom, even for the key elements like plant construction and product planning. It was all about being "different." No other company in the world would prepare a major business project in this manner.

The initial results of the Saturn planning must have pleased Tiberio, for when his two vice presidents, Alex Mair and Bob Eaton, requested a "HI-POT," possible future vice president to accompany them on a European Auto Show tour, DeKoker was selected. Executives usually give their "HI-POTs" this kind of exposure to enhance their careers.

Returning to the office from the trip, Mair expressed his and Eaton's great disappointment in DeKoker's lack of knowledge about the company's business and product. Tiberio's retirement followed soon after this. Not surprisingly, DeKoker also left. He became an executive with the Canadian auto parts manufacturer Magna International Company.[12] He was let go from Magna just a few months later. Gus Pappas' prediction about DeKoker being moved too fast and not given time to gain needed corporate experience was correct.

Outside firms frequently hire GM executives because of the corporate position they held and for the advantage of an inside business connection within GM. There are many executives who have been a great asset to their new employer, but any number could be quickly named who left GM and joined an associated outside firm only to be let go months later after the discovery that the person was not worth the money. Having the right person in the right job was not GM's policy.

DeKoker and his committee surely did their humble best and should be proud of participating in a great new enterprise. With the Saturn plant now shuttered, any wrong belongs to management for asking a group of people to do something beyond their limited experience and knowledge. Passion and "new" are not a full formula for success. There needed to be voices of experience with the freedom to give critique. This flaw in the Saturn plan will be considered in the next chapter, but the problem of inexperience and "team play" wreaked havoc in other areas.

## No Bad News Allowed

Open, honest exchanges between executives about assembly plant problems were not allowed. Vice President Richard LeFauve remarked to me and to Assistant General Counsel William Weber, "No one wants to tell [CEO] Roger Smith the Paint Plan is failing," and Vice President Robert Stempel didn't know his new assembly plants were all out of compliance because GMAD forbid the disclosure. While some companies were working to improve internal communication, at GM it was just the opposite. Everyone knew that keeping quiet was the best way to keep your job.

At one point, Roger Smith, as chairman of the board, told a group of executives to pad GM's federal fuel economy credits by claiming additional credits off vehicles made in Mexico and Canada and by Toyota in California *for* GM. There was a financial incentive to do so. The government plan had set an automotive fuel economy goal and it rewarded companies for U.S. manufacturing of vehicles that exceeded that goal and penalized those that failed to meet it. Betsy Anker-Johnson, my vice president, told Smith that his directive was illegal and asked attorney Elmer Johnson, a board member and close associate of Roger's, for support, but at the time Johnson said nothing.

A graduate of Yale University and University of Chicago Law School, Johnson came to GM in 1983 from the Chicago law firm Kirkland & Ellis. He was widely known to be critical of Smith's plans for Saturn and the North American Operations, but he undoubtedly knew the cost of speaking up. It wasn't until 1988, five years into his employment, that Elmer Johnson rebelled and spoke boldly by calling on GM to demote or dump many of the 500 managers, to form product teams with engineering and manufacturing under one management, to dump money-losing parts operations, and to slash the number of committees. It was noted in *Automotive News* Johnson played a role in eliminating 36,000 white-collar jobs at GM.[13] He also wanted to reduce the hourly labor force by 80,000. Some had considered Johnson as a possible successor to Roger Smith but that was to be Elmer's last GM meeting. He left the team and soon after, GM.

Another contender for Roger Smith's job of leading General Motors was F. Alan Smith. Alan Smith came up through the financial ranks. He was a born bean counter, sober and unapproachable, but he truly had the best interests of the company in mind. Remember my report to him in 1988? Alan saw the handwriting on the wall and more than once presented data proving that the North American operations were financially draining GM because of the Paint Plan. Finally, Roger Smith had enough and snapped, saying, "Alan, I'm tired of hearing this shit!"[14] Alan told Roger once too often that the assembly plant and paint programs were draining the company funds. Roger had staked his reputation on these programs and with that, Alan's aspirations suddenly ended, and he was soon demoted. In 1992, he was kicked off the board and retired.

The management style of two individuals, Roger Smith and Joseph Godfrey, evidenced the host of problems that result from pride and self-interest, and one of the most insidious manifestations of their pride was that no one was allowed to question or criticize. To be successful under these two men you had to be a "company man." As I have referenced before "The most intelligent thing two human beings can do is debate an issue and learn from each other." This ran counter to the corporate culture under Roger Smith and Joseph Godfrey.

## HONESTY SQUELCHED

Another humorous case of team loyalty over common sense revolved around the Fisher Body Plant in Lansing, Michigan. This plant, surrounded by residential neighborhoods, emitted over 14,000 tons of hydrocarbons into the environment each year in the 1970s. The Michigan Department of Natural Resources had offices overlooking the plant and, because of public pressure, demanded that GM abate its emissions. But how could GM add on controls and possibly establish a new regulatory standard when it was forbidden by GMAD management? Answer: Hide the fact from environmental agencies.

I learned somehow of a meeting that was to be held at the GMAD office regarding the Fisher Body plant issues. It is possible that Vice President Paul Pender had leaked it to me and wanted me there. Upon entering the room, I heard the debate on the construction and installation of a massive sheet metal ductwork on a high bridge from the paint spray systems to the powerhouse boilers for the incineration of the fumes. An engineer spoke up—it must have been for the second or third time—saying the agency should be told about the project and that the law required a permit for this device to be installed.

Pender responded forcibly, "I don't want to tell you again…no permit."

I actually felt that the engineers would be penalized if they spoke again. So I said, "Can I speak, Mr. Pender?"

"Yes, Nick," was the reply, as if my question had been expected.

I explained that the ductwork required large structural supports to hold the system high in the air. The construction would be visible to all the neighbors, and directly in view of the agency's offices. The construction workers would talk to the plant employees, who live right across from the plant, about what they were doing. The agency had friends who worked in the plant and knew everything that was going on. I told Pender that the plant could face serious fines for construction without a permit.

Mr. Pender stopped me and said, "It's almost 4:00. Call the Michigan agency before they go home and tell them we are applying for a permit tomorrow." My statements allowed him to do what he probably knew was the right thing all along, but could not do without displeasing Godfrey. He seemed relieved that I had forced him.

I experienced the pressure of this culture first hand. Personally, I was disinterested in the game. I was taught to fight every day for my company to be profitable such that every employee and stockholder is rewarded, not just the few in power. Privately, I did a study and in 1965 published a report to management that funds for powerhouse boilers and air compressors were being appropriated using poor engineering calculations and assumptions.

The powerhouse chiefs were assuming extreme outside temperatures and demand conditions that never occurred. Therefore they were encouraging management to overbuild. They would not accept a drop in boiler pressure, usually over 200 psia (pounds per square inch absolute) even for a few hours, when *production used only 15 psia.* I know from personal observation that there isn't a GMAD plant that uses one-third of the electrical power, steam, or compressed air available to it. All the powerhouses were overbuilt. At the time, my report was ignored. I was told by Ken Robinson, GM's ventilation and heating expert, that an effort was made by powerhouse operators to have me fired over my report. In particular, Gail Kavanagh, the Kansas City powerhouse chief, went after my job because I questioned his request for a third coal-fired boiler, despite the fact that his plant had never even fully utilized the output of one.

Time proved the veracity of what I was trying to communicate. The Fort Wayne truck plant has two very large and expensively built coal-fired boilers, each capable of producing 200,000 pounds of steam per hour. In actual production, neither of these boilers was used because a single 100,000 pound per hour gas-fired boiler could maintain the plant's needs. Today outside contractors operate GM powerhouses at operating usages well below 50%, with some plants even below 80%, of the requirement predicted by those operators that opposed me.

While I was speaking out and getting labeled for not being a "team player," it seemed to me that everyone else was ruled by an instinct for self-preservation instead of a desire to do what would be best for GM. Being so successful and number one in the auto industry had once given all employees an unusual sense of pride of being part of the best. Workers were proud to tell friends and anyone they met "I work for GM." However, during Roger Smith's tenure as chairman of the board, the company continually lost market share and suffered billions of dollars in losses.

Roger Smith asserted himself into discussions and asked the right questions whenever he attended a meeting. He wrote answers to his questions down in a red schoolbook, and had the goods on a few fellows lying to him, but never seemed to do anything to one of his team players. He and others made most employees feel they were simply employees, and no longer part of a team or family. Was taking care of the employees a high priority? Under Roger Smith's helm, salaried and hourly workers alike felt threatened and unappreciated.

The welfare of the employees was nowhere to be found on the priority list. The spirit that had bound us together evaporated. Other executives agreed, the greatest harm Roger Smith did to this fine corporation was losing the employees' "esprit de corps." It was replaced by this spirit of self-preservation.

I personally observed the out workings of self-preservation in a management meeting in 1985. Of the five executives in the room, I was the lowest ranking, attending at the invitation of the GM President James McDonald. Joining us were Executive VP Lloyd Reuss, and from the Truck and Bus Group: VP and Group Executive Charles Katko, and under him, VP and Group Director Pat Coletta. Reuss proposed to McDonald that he be permitted to convert the 75-year-old Fisher Body, multi-story plant into a second "Buick City" plant. The first Buick City was touted for its synergy in producing most of the vehicle parts in buildings adjacent to the assembly plant. Reuss proposed to McDonald that the vehicles produced at the new plant would be of the highest quality, made at the lowest cost, in the fewest labor hours.

After Reuss promised all that anyone in the American industry could hope for (but never deliver), McDonald thanked Reuss for the great presentation but said, "I must close plants. We have too many, and taking a 75-year-old plant apart to remake it as an assembly plant is out of the question." Then McDonald turned to Coletta and with a fierce look and loud voice yelled, "Why can't you do what Reuss can do?" and he and Reuss walked out of the room.

Coletta, red-faced, yelled at Katko, "Why didn't you defend me? Reuss can't do any of those things in his plants. Reuss is full of it!" And again to Katko, "Why didn't you speak up?" But Katko could not question McDonald for fear that he would get hurt himself. The culture of "team play" dictated that honesty be squelched.

As a footnote, the second Buick City was never built, and the original one, with its miles of space and hundreds of buildings, is gone today. Reuss was a good talker with little substance.

## Wisdom Rejected

In the midst of this, there was one executive who understood the company culture and the harm it was causing. He could define it in clear, easily understood terms, and knew what must change. Today, ask any member of management what is meant by the "culture," which outside experts claim must change, and you'll most likely get a vague and meaningless answer. Vice President and Chief Economist George C. Eads risked everything to save GM, but was forced out because his approach to the major issues facing the company was in such variance with the approach of the rest of management.

George C. Eads is a remarkable man who tried to make a real difference in how GM operated, but was not considered to be on the "team" and was turned away. His most radical recommendation was for GM to stop investing in new plants and to buy the Toyota Company. At the time, Toyota had 4% of the market and produced $4 billion in profits each year in the 1970s. In 2006 Toyota made $12 billion in profits and in 2007 had over 12% of the market.

Most telling of the differences in philosophy between Eads and the other vice presidents was what came out in a startling fashion at a round table meeting held for staff directors to keep them informed, and to encourage us all to stay the course. Five vice presidents, representing the five key management departments, sat in the front of the room and addressed six major issues confronting the company, including how GM should approach the EPA and government officials on safety and emission issues. Each gave a short response regarding what they would do. George Eads was always the last to speak on each issue. He spoke with conviction, presenting a distinct approach that seemed to be more sensible, logical, and reasonable in comparison to the common themes of the others. For instance, in regard to the company's response to an EPA proposed regulation, the first four VPs would not volunteer any cooperation. They all wanted an even playing field with the other auto manufacturers, or a law, before doing something. Eads proposed cooperating with the EPA, to the extent that the solution was practical and cost effective, and in return, receiving from the EPA extensions or postponements for controls too costly or not cost effective.

Since the first environmental and safety acts were passed in the early '70s, GM has followed a pattern of resisting every regulation, but eventually installing the equipment, then bragging and advertising their environmental advances. Innovative giant, the 3M Company, responded in a positive way to the EPA, and their bottom line shows the success of this strategy. In my field of plant pollution control, I was accused of loving the agency more than the auto industry, but it was precisely because of my relationship with the agencies that none of my requests for relief were denied and none of my plants were fined. Even Chrysler and Ford managers knew I could get relief that they could not.

Eads was absolutely right in his approach on every issue presented at that meeting. I will bet my reputation, career, and pension that George Eads would be of eminent value to the new GM management—but I doubt they would listen. His letter proposing the purchase of Toyota in lieu of the plant modernization programs could have saved the company from bankruptcy. Additionally, his letters to management recommending that they stop saying, "We have turned the corner," were truthful and well written.

While walking through the Detroit Auto Show one January, Eads said to me, "They don't want me."

I replied, "Same here." Neither one of us was a so-called "team player."

# 6

# Unchecked Mismanagement

Computer industry legends like Bill Gates and Steve Jobs were known to have aggressive, hands-on leadership styles. They knew the consumer, they knew every intricacy of the product, and they were relentless in their pursuit of excellence at every level. Jobs was a fanatic about user experience and the design of products. Gates is a fanatic about running the engineering groups and upholding the *quality* of them. These men were very successful because they were fanatics about the most important things in their business. GM needs leaders who live and breathe vehicle assembly and are never satisfied with any plant's performance.

The same problems that existed at the board level were rampant at the management and operations level as well. In the absence of meaningful knowledge of the business, pride and self-interest permeated GM at the management level. Their presence resulted in business decisions so costly to GM that the folly of these business decisions must be considered in any conversation regarding the downfall of the once mighty corporation. The successful history of GM generated a dangerous level of pride at the management level. The financial staff, in the good years, the '50s and early '60s, had what seemed to be a surplus of money. Millions were spent on expanding the plants, their powerhouses, electrical distribution systems, and warehousing, without much justification. The company, once known for sound financial planning and stringent cost controls, gradually became complacent and there was not enough knowledgeable oversight to catch and correct poor decision-making.

## No Cost Control on Construction

In 1968, my supervisor, Jack Cragen of the Chevrolet Central Office (CCO), requested a review by the financial staff of the CCO procedures for processing appropriation requests for plant construction projects. A tough lecture on cost controls was expected from the financial staff. Instead the response was, "Isn't it great you work for a company where you don't have to decide whether your project A or B should be approved, because both can be approved." We couldn't believe this response was not a joke.

Management and the car divisions were on a roll, so why question their judgment? This is the point at which corporate pride and personal self-interest linked arms. CEO Roger Smith was from the financial staff, so when he wanted the new assembly plants, all with robots, and he wanted the all-new Saturn complex, there was no serious debate of the proposals by the staffs. As shown in the last chapter, the decisions of the chief executive officer at the mighty GM could not be questioned, and those that did criticize or question soon suffered for speaking up.

I've often questioned why there isn't an independent group of "old hand" experts, beholden to no one except the company and the stockholders, to review GMAD projects. So many major capital outlays for plant construction or for changes to plant utilization were never seriously thought through and billions of dollars have been wasted as a result. Even recently, GM was considering construction of a new building for production of the Volt battery until bankruptcy forced the use of an existing building. The new building would have added an up-front capital cost as well as an ongoing shipping expense to take the batteries to the cars.

Marketing and the financial staff have consistently given in to the boss. Vice Chairman of Global Product Development Robert A. "Bob" Lutz, championed some of the best vehicles in the market, but also many that were losing propositions from the start. He seemed uncontested in promoting specialty "niche" cars that increased short-term revenues but generated long-term losses. He modified the back of a popular SUV so that another division could sell one like it, but this resulted in stolen sales from the first division. Lutz was a top "car guy," but even a product manager with a great reputation needs to have meaningful control so that costly mistakes are avoided. No one stopped Lutz from spending billions in an attempt to revive the Saturn brand after its assembly plant was first shut down. His plan called for the production of Saturn models at other GM plants, but after a few years, the revival attempt failed as well and the retooling investment was lost.

The Honda board reviews all new product proposals and demands that firm design goals be met; including miles per gallon, vehicle weight, production cost, and sales numbers. Honda's leaders are experienced enough to ask all the right questions and to insure that the product manager hasn't overlooked anything. If a member's proposed product misses a target, that is the last product he'll ever offer; he is usually fired. Lutz was allowed to produce many more products after his initial ones accumulated losses. Where is the discipline in the management system?

## No Internal Review

As an example of what I witnessed, the GMAD Wentzville assembly plant wanted me to contact the EPA and request that they be granted more time to issue a powerhouse emission report that was already far overdue. I went to St. Louis and, walking through the production area late at night, noticed a lot of cars being repainted. The plant engineer found that oil and grease from the overhead conveyers were spotting the vehicles below, so I felt assured that a fix was underway. I made excuses for the plant and got a three-month delay for a report my people could have prepared in a day. Months later, not having received the emission report, the EPA was ready to penalize the plant for violation of the law, so back to the plant I went. To my amazement, vehicles were still being repainted because of the same problem, and the plant engineers simply laughed about it. The plant manager and I flew to EPA's regional offices in Kansas City, promised to hire a contractor to finish the report and submit it within the month. Thirty days later no report was forthcoming so I told the EPA to fine the plant—I gave up. With an unrepaired quality issue and an EPA fine, was the plant manager fined or fired? No, the plant manager became the vice president of Global Manufacturing and Labor Relations.

GM has been assembling vehicles well over 100 years yet surprisingly, when compared with other auto manufacturers, it has many of the least efficient assembly operations and highest operating costs in the business. Joseph Godfrey's Oklahoma City assembly plant, which had been constructed to Godfrey's standards while he led the Assembly Division (GMAD), was reportedly sold in 2009 to the government for $70 million after *hundreds of millions* were spent a few years prior to improve its operations. Five years prior, this assembly plant was to be closed under Wagoner's reconstruction plans because out of all the GM plants it had the poorest use of manufacturing technologies. How did President Pete Estes allow Joe Godfrey to build a plant without the latest manufacturing systems? How did McDonald and Smith allow so many oversized new plants to be built? Management, like the board, knew very little about the actual manufacturing of its products!

The GMAD system planned four assembly plants, two to build the two-door Beretta and two to build the four-door Corsica. Marketing predicted these very similar models were going to steal the market, but as they rolled onto the dealer lots they proved to be underpowered and cheap looking. Sales numbers indicated that only one or maybe two plants were needed to meet the demand. When asked to build the two-door and four-door version of the same model on the same line, GMAD reported that the sheet metal body shop and other areas of the plant would require prohibitive reconstruction. As a result, two plants were left to operate at very low capacity and at a loss. Contrast this with the Ford truck plants equipped to build six different *models* on the same line. With its years of experience, GM should have been a leader in innovation, not a victim of changing times. But, as I stated earlier, there seemed to be no gathering of these experiences, no measurement of each plant's performance against a set of goals, no exchange of the best practices between plants—in other words, no management. It is true that up until Roger Smith's corporate restructuring in 1984, GM had been comprised of several independent car lines that often continued to compete against one another, but supposedly we had moved past all that.

GMAD has made so many profound mistakes over the years in managing assembly plant projects that common sense dictates a need for a higher body of experts to review their projects. Their rubber-stamping has to end. History will look back at President Elliott M. Estes, President McDonald, CEO Jack Smith, Robert Stempel, and Lloyd Reuss as having taken promises from GMAD without questioning its many wasted expenditures. From top to bottom, quality was assumed, not pursued, and the ability to throw large quantities of capital at any need was pridefully practiced. There were no internal rewards for quality or for cost containment in construction. And when reviews finally did come, they were often late in the process.

In a vivid example of upper management's failure, President Lloyd Reuss came to discover one major deception after it was too late, when a related deception came across his desk. In the fallout, a good vice president paid the price for going along with the internal culture.

Not long after the 1987 Paint Plan end date passed, an appropriation requesting $33 million for pollution abatement controls on the Janesville assembly plant was submitted. It came from Pat Coletta, VP of the Truck Group, with the justification that the law demanded more pollution control. When the appropriation reached the desk of President Lloyd Reuss, he immediately called my vice president, Betsy Ancker-Johnson, and demanded, "Why do I have you on staff if you can't save me from this?" He rightly felt that all of the additional costs listed for pollution control should have been addressed.

Ancker-Johnson rushed to my office, showed me the appropriation and asked if the money was required. I knew right away that it was a ridiculous request (remembering the Godfrey edict: no abatement) and I replied, "None of this is required and I never saw this appropriation." More than once, my employees were pressured by GMAD to bypass me, but in this case, it was a vice president in the truck group who personally persuaded Greg Peters, a staff engineer, to approve the appropriation, both knowing it was all a lie.

So if there wasn't a legitimate need for additional pollution controls, why was the Janesville, Wisconsin, plant project short of money by as much as $33 million? The truth was that months before, a choice had to be made for a new truck to be produced at either the Flint truck plant in Michigan, or the Janesville plant in Wisconsin. As in any plant selection, consideration must be given to both the actual cost of production and the expense of shipping. The parts must be shipped in, and the product must be shipped out to the buyers, therefore the location of the plant relative to its suppliers and its market is critical. Normal business practice naturally dictates that the plant with the lowest production and shipping cost should be chosen. The difference in the expected cost of production and shipping typically varies by $100 million per year from plant to plant. Accounting estimates had indicated that it would be $400 per vehicle more profitable to produce the new truck in Flint. Multiply that number times the expected production of 300,000 vehicles annually. For the Janesville plant, Lake Michigan created a shipping obstacle between the plant and both the parts suppliers and the final market on the East Coast. However, Coletta and others wanted the project at Janesville. Consequently, they fudged the numbers and submitted an appropriation favorable for the Janesville plant. No one in upper management knew of the inaccuracy. The top executives trusted their VPs and never checked the figures.

As the project got underway, the shortages appeared immediately. An appropriation was written for $33 million with the justification that it was needed to "meet unforeseen environmental regulations." Despite the approval signatures of three engineers and five executives from the Truck and Bus division, I staked my reputation and job on the belief that there was no pollution control planned whatsoever. The whole appropriation was a ruse. I knew that Coletta always followed Godfrey's dictates of "no control or else," and this gave me a great deal of confidence. Looking at the document in my files today, I have to wonder if all eight people who signed off on the request were involved in the deception, or was this just another example of poor management.

To give me support in the meeting dealing directly with the vice president, my director, Dave Eden, had a member of the financial staff join me. He knew this was a serious situation. We met in the VP conference room and an engineer tried to examine each item on the appropriation. As each line item was shown on the screen, my most frequent comment was "I can get you out of that requirement." Only for a few line items to the tune of $2 to $3 million, cited as a need for new spray guns and better robots, did I give my approval.

I said that we supported those items because they had the potential to lower production cost and improve paint quality. Coletta and I both knew that those items had been included in the original appropriation for the product production and were bogus. We also knew that if approved, none of the money would be spent on the listed pollution controls.

At the end of the meeting, Coletta thanked me. In the hall, the man from the financial department said to me, "Thank God, you gave him something to save face." But three weeks later, I got a call saying Coletta was found crying in his office because Reuss told him to retire. The weight of the Paint Plan on the corporation caused good men to be dishonest and the management culture had no structure to preempt the dishonesty.

The truck division also used the exaggerated costs of pollution controls as an argument to avoid applying them. A letter to the state agency stated that pollution control devices would require ten men per shift, three shifts per day, to operate and also ten additional maintenance employees. The truth was that no additional personnel would be required after the controls were installed. Terrible games were played with appropriations during the Paint Plan. Today the Janesville plant is closed and the Flint plant is being expanded because it has a cost advantage.

Throughout the corporation, lack of oversight allowed self-interest to dictate decisions. Beginning in the 1970s, GMAD expanded the assembly plant facilities like the powerhouse boilers and compressors, the electrical distribution systems, pollution control for waste treatment, and even the parking lots with the thinking, "We get the money, others won't." It was all about self-interest. They operated under the assumption that large capital investments would protect them from future closures. Though self-serving, the assumption seemed reasonable. The truth: many plants were closed a few years later and the capital expenditures were never recouped. Appropriation requests were blindly approved. In my years at GM, I never saw one risk analysis demonstrating that the existing systems were adequate to meet production needs—though generally they were. Is there any evidence of financial studies justifying these facility expansions? As you will recall, my 1965 report to Chevrolet management about the overbuilding of powerhouse equipment resulted in an effort to have me fired by the powerhouse chiefs.

In addition to the expense of excess power capacities, GM found itself in a position of excess laborers. The union/company contract allowed employees losing their job to receive 90% of their normal pay while waiting for GM re-employment in what became known as the "body bank." It was a very generous offer by management under the assumption that business would improve in the not too distant future. Unfortunately, improvement did not occur, yet no one thought to consider the corporate cost or modify the contract.

Many body bank employees were displaced by expensive computer-controlled welders, robotic painters, and a host of "laborsaving" devices intended to improve productivity and reduce labor costs. But management lost much of the productivity benefit due to the increased costs of the body bank. Some plants, having installed laborsaving devices, kept the workers in the plant because managers didn't want to pay employees to sit at home and do nothing. When this happened, GMAD had no way to accurately measure improvement in a plant's bottom line, even after spending millions of dollars to upgrade the production system with new technologies.

Was anyone responsible for tracking the number of employees or man-hours, to produce a vehicle before and after a major modernization project? Were plant efficiencies before and after compared to support the claim that new equipment would reduce the workforce? The only report I know of compares vehicles produced per hourly employee for Ford and GM and shows GM hovering around 12 vehicles per employee throughout the early '80s, while Ford rises from 11 to 16. Meanwhile the Japanese competitors are approaching 100 vehicles per employee.

The plants should have been ordered to eliminate as many workers as possible, and to reward plant managers who used the fewest workers. Robotics that increased the labor pool should have been flagged as inefficient. Perhaps the body bank problem would have been solved if the plants were ordered to send all surplus employees over to the lobby of the GM headquarters. This would have forced GM management to address the problem!

The bottom line is that the GM manufacturing culture never concentrated on making the assembly plant operations the best in the world. An extremely accurate statistical analysis of every assembly plant's productivity should have been instituted and before being undertaken, modernization projects should have been assessed for their long-term benefit to the corporation. However, GM manufacturing plants had a greed of their own.

## No Corrections

Hydromatic Division transmissions had been failing in the field for five years due to a roller pin bearing that slipped out of place. The result was high warranty costs and the assembly of hundreds of thousands of replacement units for the customers. A few years of this finally got the attention of GM President James McDonald (1981–1987), who decided to investigate and find a solution. He sought the advice of Ray Hoste, an honest and highly qualified staff engineer assigned to this issue. After an explanation of the problem, McDonald requested a sample of the failing parts to take to the Willow Run plant.

"Look on the sides of the expressway to the plant for the parts," replied Hoste.

"Why?" McDonald asked.

"Not too long ago, the plant production superintendent, with a demeaning 'do you hear me,' ordered us to place a box of parts that failed, not in the trunk, but on the floor in front of the passenger seat where he would be sitting, presumably so the parts could be thrown out of the car window." Hoste was a great engineering mechanic who had the knowledge and desire to fix the problem. But, he was effectively stopped. Why? Self-interest.

The GM bonus award system is based on plant revenue/profits and units produced, with little or no penalty for unit failures in the field. In fact, parts replacement and repair were perceived to increase revenue! The short-term profit was better if the problem was *not* addressed. This insane bonus system placed the plants' short-term profits over the corporation's long-term interests. Where were the GM transmission experts to correct this problem? Why did corporate leadership not insist that the plant managers secure long-term profitability by looking out for the customer? Self-interest reigned at all levels in GM.

President McDonald was frustrated by the poor product quality and undoubtedly weary of the negative press reports, but the culture was not easily changed. In addition to the concerns cited above, many of the complaints concerned the "fit and finish" of the vehicle. When separate body panels meet on a well-built vehicle, the panels are "flush" and the gap between them minimal. A lack of attention to these details cheapens the appearance of the vehicle. Complaints were common throughout the '60s and into the '70s concerning the construction of the body and the rust protection—or lack thereof. When these problems were allowed to persist, the public discerned that "the company simply doesn't care."

During that time period, a friend complained that the air noise from around the driver's door overpowered his ability to hear the radio. Fisher Body's Grand Blanc, Michigan, press plant engineers told me the door dies had been redressed (ground smooth) so many times in lieu of being replaced that the door seals no longer sealed properly. His Cadillac had two seals, which I adjusted with little reduction in noise levels. Cadillac models the following year were advertised as having three door seals, a great announcement with Japanese models boasting just one seal and little air noise. When the dies to make the doors were finally replaced, the problem disappeared but by then the quality reputation had been long damaged.

GM's lack of concern for the quality of the vehicle's appearance in the '60s and '70s was also evident on cars with rusted top surfaces (hood, roof, and trunk lid). Most likely, these cars were caught and held for too long in one of the cleaning, painting, or oven stages in the paint shop because of a production-line stoppage. They should have been reprocessed for proper coating.

In another example taken from the mid-'70s, the Saginaw Steering Gear plant, under the leadership of Joseph Godfrey, produced units with hydraulic O-ring seals that shrank and leaked on cold morning start-ups, effectively disabling the steering power assist mechanism until the engine compartment warmed up. It became known as "GM's morning sickness." Was the problem addressed at the production level? Not for years! Finally, in a redesign of these units, it took a courageous young engineer (a friend of mine) to get approval to grind the O-ring slot in lieu of the standard machining of the slot. This manufacturing redesign got the problem solved once and for all, but only after hundreds of thousands of units were out on the market and had to be replaced. And while many a mechanic was finding financial gain buying and installing repair kits, GM's quality reputation was undercut and warranty costs escalated. Yet to my knowledge, no one at this plant under Godfrey was penalized for letting the problem go so long unchecked.

Perhaps it was an unspoken awareness of these unhealthy ingrained attitudes that led Roger Smith to be so eager to embrace all that was "new." Ironically, the pursuit of the latest innovations depleted the very funds that could have been spent to address the quality concerns held by consumers. Instead of looking to the past for lessons that could be applied to the future, GM, under the leadership of Roger Smith, forged ahead with great enthusiasm and fanfare into uncharted waters of developing technology. This is part of the mindset that allowed the Paint Plan to make promises it could not keep. The latter part of the book will finally shed light on the magnitude of this financial drain, but the same flaws in thinking were at play in a more public arena, and you can see the result when you drive by the partially shuttered Saturn plant in Spring Hill, Tennessee. Saturn was a program with great promise and public fanfare. The publicity was great, but the product was not. Numerous failures in implementation led to its eventual demise.

# 7

# The Culture of Resistance

One of the manifestations of the attitude of GM pride combined with ignorance was the culture of resistance to government regulation, particularly in regard to vehicle safety and protection of the environment. The culture of resistance was so strong and ingrained that almost no one, except for Eads, recognized the financial costs and public relations costs that would mount over time. These were significant costs that could have been easily avoided with compliance. In a day when industries are competing for the title of safest, cleanest, and "greenest," it's hard to fathom the lengths to which GM went to avoid installing safety devices and to protect its corporate right to perpetuate practices that were unquestionably contributing to global pollution. No one thought to ask, "What is best for the customer—current and future?"

In 1970 Congress enacted amendments to the Clean Air Act with imposed compliance dates and auto emission standards. This act forced the companies to build and install catalytic converters on all cars and trucks. In 1972, after the Environmental Protection Agency (EPA) was established, GM President Ed Cole hosted a dinner meeting, supposedly to offer a hand of cooperation to the EPA, the Michigan Department of Natural Resources, leaders of environmental organizations, and the press.

With the vice president of Industry-Government Relations Staff, Robert F. Magill, pulling on his coat sleeve, saying, "Don't say it," Cole, sitting just two seats away from me, nevertheless said, "If you force us to install catalytic converters, we will probably go out of business." The resistance had begun and the ignorance of the board and upper management in the areas of environmental and vehicle safety was at times appalling.

## Rooted in Ignorance

Vice Chairman of the GM Board, Howard Kehrl, a Phi Beta Kappa honor society member, seemed not to realize that flushing acid down the drain was against the law. On one occasion he asked me, "What is wrong with having acid rain?" He continued: "I have a mansion with copper gutters cleaned with acid and they really shine."

Surprised, I responded, "Acid rain attacks human lungs, turns lakes acidic causing the fish to die, eats away at statues and structures, etc." He left saying no more.

And while the board exposed their ignorance of environmental issues, the Environmental Activity Staff (EAS) was overstepping its bounds into the area of vehicle safety. Thinking management would be pleased, the executive director of the Environmental Activity Staff recommended to the management committee that GM could save over $15 per vehicle, by *not* offering four-wheel antilock brakes. With this recommendation coming from the EAS staff, President Robert Stempel was surprised, and said, "How could GM be the only company without four-wheel antilock brakes?" But ignorance in regard to braking systems extended shockingly to the level of the director of automotive safety for GM. Following a director who had fought unsuccessfully against airbags, the former assistant, now director, told me with enthusiasm of his experience repairing a church bus. He was intrigued by the process because it was the *first time* he had seen and replaced front brakes. I was shocked by his lack of expertise, but why was I the only one? Had my career path given me a unique perspective?

When my education in automotive engineering began at age 13, it was not in a classroom, but on the factory floor. I had learned how to operate most metal cutting machines and had gained an understanding of their potential and their limitations. I had gained hands on experience with stick shift transmissions and I had rebuilt front and rear brakes on several different models. With this background, when I went on a trip to Japan with GM, I was quick to observe that that Toyota's safety director knew more about brakes and brake systems than their producers and repair mechanics, and certainly more than the safety director at GM. There is no question that knowledge makes a difference in Toyota's approach to safety. But at GM, resistance to regulation was a higher priority than consumer safety, so product knowledge was undervalued. The management message was clear: "fight regulation."

## Resistance to Safety Regulation

After the National Highway Traffic Safety Administration (NHSTA) mandated the use of seatbelts and soon after, the use of airbags, the first vice president of the EAS made an initial announcement that it was against the U.S. Constitution for us to force our customers to wear seatbelts—so why install them? The second EAS vice president's opening pronouncement was—GM has done enough for vehicle safety; there is no need to install airbags. Neither man was a safety expert.

Elmer Johnson, newly hired head of the GM legal staff, with no auto safety background, had lunch with us and during the meal he said, "We don't have to install airbags; we've done enough."

He was a lawyer, not a safety expert! In hindsight, I wondered if he was hired just for this purpose.

He continued, saying, "We'll get someone the public trusts to say, 'We've done enough.'" When asked whom he suggested, Johnson responded, "Walter Cronkite is the most trusted man in America; we'll get him." When it was pointed out that Cronkite didn't know anything about automotive safety, Johnson answered, "We'll write for him and pay him for his efforts."

Did anyone really believe that the revered TV news anchor would accept the offer? This wasn't breakfast cereal they were asking him to endorse. This was a matter of product safety—the difference between life and death in an auto accident. As I listened dumbfounded to this lunchtime exchange, I had to wonder, had GM been successful with this tactic before?

GM had a staff of "Societal Risk Assessment" experts in the GM Research Laboratories to support management's efforts to avoid safety regulations and not to propose that GM do more. In 1979 they held a symposium, "How Safe is Enough?" with experts in the field representing diverse views on the subject. The GM presenters offered evidence that the average person is not capable of analyzing data and making a proper decision about risks they may face. GM, on the other hand, could because of their expertise. To counter GM and to make the opposing point that the public at large makes the best decisions about their own lives, a Mt. Everest climber, Michael Thompson, presented two completely different lifestyles in Himalayan villages a few thousand feet apart in elevation. Life at the different elevations presented different challenges. People, he stated, adjust to the environmental and economic land conditions nicely without others telling them what is right or wrong. Later, a GM speaker attacked Thompson's ideas about risk takers and who is best qualified to decide important issues. To me, it seemed defensive and in bad taste. The symposium's final speaker, whose role was to summarize the conference, skillfully tore into GM's arrogance and self-righteousness. It surprised me that he would do such a thing to the symposium's sponsor. Months later, GM published the book of the proceedings but replaced the final talk with another one of the man's speeches. I had to wonder about the ethics of this.

Betsy Anker-Johnson, the third vice president of the EAS staff in seven years, told me, "If people knew what I know about how dangerous steering wheel airbags are to some people, they wouldn't want GM to install them." Engineering staff told the vice president that GM could be sued over injuries and deaths caused by airbag expansion, which violently threw back the driver's head, particularly if the driver was small and older and hunched over the steering wheel. Did the vice president inquire if there was an easy solution to this problem? She did not, but not long after that, safeguards were added and airbags were installed industry wide, making their use one the most important safety features in a car today.

Who at that time would have believed that Chevrolet would offer a 2012 vehicle with 10 airbags? It would not have been David Martin, director of Automotive Safety for GM. Under his leadership the company pursued all of the Michigan congressional representatives, the best lobby team in Washington DC and the Business Round Table—anything to delay the requirement of airbags. Only when Chrysler announced in 1978 that it would install air bags on all their vehicles did GM give up the fight. That very afternoon, upon learning of Chrysler's capitulation, the Safety director announced his retirement. The battle was over! All of these important executives were obviously hired by management to resist further safety regulations and the same resistance was expected for automotive emissions regulations. All were brilliant people who had the brains to ask the right questions and offer some opposing views. But they didn't.

## Resistance to Pollution Regulation

This same EAS vice president received a recommendation from the Automotive Emission Group director that GM offer to install the carburetor hydrocarbon adsorption canister proposed by the EPA. He felt that what was now an EPA proposal would soon be mandated, and in fact, the installation could be easily accomplished at a low cost. In return for cooperation, GM could ask the EPA to postpone any regulation that was too costly. But the idea of cooperating with the EPA didn't go over too well with the vice president, and the director was transferred out of the group. A few months later the addition of a canister became the law.

It was in 1973, during a period when the assembly plants were temporarily closed due to the oil embargo and the small parts plants were being eliminated, that Gabe Tiberio, director of Plant Environment for the Environmental Activities staff, was commissioned to form a corporate energy group to address the energy problem. It was at this time that the largest shift in attitude towards environmental regulation occurred within GM. The Divisional Central Offices adopted the practice of impressing management by resisting further regulation. Some of my best people stopped aggressively solving pollution problems and began resisting them at all costs, *even if it was better to solve the problem and thereby save or even make money for the corporation.* There were constant battles with the EPA during the implementation of the Paint Plan. While its intended public purpose was to reduce air pollution, it gave rise to the most costly examples of EPA resistance, and in its implementation, GM went to the most ridiculous extremes to mask repeated air quality failures. The Plan was by far the most outrageous commitment ever made by any company. Many in top management knew it was full of false promises, but could not stop the program once it had been presented to the EPA.

After the 1973 oil embargo, the theme resonating through the company was "we can't deal with more regulations and non-product expenses." With so many new federal regulations, management's frustration grew, sending the wrong message to engineers regarding the approach to problems. One executive, for example, Executive Vice President Richard L. Terrell, concerned that the proposed EPA's tailpipe emission standard for reducing hydrocarbons from 95% reduction to 97%, approached engineers in this manner: "We can't meet this standard, right? There is not enough platinum in the world, right? We can't guarantee the catalyst's performance for 100,000 miles, right?" I was there in the meeting as the executive spoke authoritatively and in a firm voice; he wasn't seeking the facts, instead he was giving orders. There were some in management who would have asked, "Can we meet this new requirement and what will it cost?" Or better yet, "Is it possible to make money by doing so and simultaneously improve GM's environmental reputation?" The answer to all these later questions was yes, yes, and yes. GM charged $400 for the first catalyst and the aftermarket sold them for $45.

## The Financial Folly of Resistance

A PhD chemical engineer from my Water Pollution Group, was given the responsibility for the Waste Management Group. This very bright fellow devoted most of his time, even weekends, to a well-intended effort to save GM the expense of government-mandated controls. He searched through the laws and wrote copious letters for the plants and GM lawyers with interpretations of the law on ways to avoid "costly" solutions. He soon developed a great reputation with the plants and Divisional Central Offices. But, according to his reputation at the EPA, he did not solve problems (per a former EPA administrator). The worst and best example of the engineer's approach was his handling of the Massena Foundry PCB problem. According to the EPA, polychlorinated biphenyls (PCBs) are synthetic chemicals, which have been demonstrated to cause a variety of adverse health effects. They are no longer produced in the United States, but are still found in the environment. PCBs had been used as coolants and lubricants in transformers, capacitors, and other electrical equipment because they don't burn easily and make good insulators.

Massena is a foundry located next to an Indian reservation, the Thousand Islands, and adjacent to the St. Lawrence River, which is classified as a pristine national treasure. Until 1980, polychlorinated biphenyls (PCBs) were a component of hydraulic fluids used in die-casting machines at the facility. Less than one hundred feet from these sensitive areas, GM had two large storage lagoons filled with PCB-contaminated fluids from a waste treatment system of low efficiency, several deep wells filled years ago with PCB, and one of the country's largest landfills filled with PCB-contaminated sand. The sediment on the river bottom and the plant's discharge pipe were also heavy with PCBs.

The Resource and Conservation Recovery Act of 1976 formalized regulation of the discharge, handling, and disposal of PCBs. In 1979 their manufacture was entirely banned. GM should have removed all of the waste to the proposed EPA standards before the Act took effect, and could have done so without much regulatory oversight and for a few million dollars.

I asked GM's Argonaut Engineering Division to estimate the worst case cost of moving all of the landfill waste. Their cost was $900,000 to dispose of all the solid wastes to EPA-proposed standards. In March of 1976, with an additional $300,000 to insure sufficient funding to clean-up all PCB contaminated soil, I presented a $1.2 million proposal to Bill Johnson, EAS director, but the Central Foundry Office engineer, Bill Cheek, insisted that new technology might someday avoid this expenditure, so he refused to authorize the removal. The hope was that a PCB-eating bug would be found, allowing the waste to be eaten in situ. It was never found.

Using the information in my proposal, could we have accomplished the removal before 1978? Yes. In fact, in 1974 and 1975, Chevrolet took most of the 100% PCB fluids from the machines to an incinerator at a cost of a few cents per gallon. Today, it would cost tens of millions of dollars. Since 1978, the two lagoons have been removed, the waste treatment system has been upgraded with carbon adsorption filters, the river bottom has been dredged and a leachate system has been put into use to capture PCBs in the groundwater before they reach the river. In 1987, a National Geographic television program on the ten worst environmental problems in the U.S. stated that GM had spent $100 million so far and had reduced leakage to the river by only 30%. Bill McFarland, GM's Land Remediation manager, told me that the figure of $100 million since 1978 is fairly accurate, but that the leakage is well under control.

For years, I predicted that the remaining waste at Massena would be removed someday. Finally, in 2008, GM announced that the Massena plant would be closed and that the final clean-up of the site would begin. I am sure that this plant had been kept in operation, even it was not cost effective, just to avoid the millions in clean-up cost that come with closure. By 2008, the price tag was much higher and the corporate coffers much lower. Even in 2000, a news report stated that the state and community oversight committee reviewing the Massena problem wanted another $220 million for additional PCB control while the federal EPA thought $60 million more might do the job. After 32 years of delays and piecemeal cleanup, I am told by inside sources that the company spent $330 million to make the property environmentally safe. Remember that in 1976 the "worst case" estimate was $1.2 million. In a parallel situation, $220 million was spent for clean up at the Bedford, Indiana, casting plant. There is not an environmental problem whose solution could have produced better results for the corporation had it been implemented sooner rather than later.

Over time, this approach proved to be exponentially more costly to GM than prompt cooperation would have been. Ironically, engineers from the plants and lawyers from the legal staff would take the letters written by the assistant director of Waste Management, use them as is or personalize them, and then show their supervisors how they were saving the company more pain by protesting against regulation. Everyone in this process loved and admired this man for his articulation of their "cause," especially his own management and the divisional central office people.

Most environmental engineers in plant pollution control, automotive pollution, or safety, wrote letters of protest to various governmental agencies during their careers. The letters addressed the proposed government regulation or new restrictive plant environmental or automotive standard. Declarations were made that the new standard was not needed, was not cost effective, didn't apply to the company, was not obtainable with current technology, or as many letters stated, if imposed they might drive the company out of business. The big three American car companies (GM, Ford, and Chrysler) have thousands of these letters in their files. Not one letter will be found stating "great idea" or "let's get at it because the company really cares about the public good." Only the Industry-Government Relations Staff knew better and felt the public wrath, political backlash, the EPA and state agencies' irritation with GM, and the harm to the company's reputation.

Retirees from the auto companies take pride in their careers and in the letters they wrote on the company's behalf. But how many have reflected on the long-range outcome of their letters? One honest man did. At a retirement party, I ran into a General Motors staff scientist who spent years researching, conducting studies, and writing letters relating to the company's contributions to our nation's air quality problems.

Just to start the conversation, I asked, "What are the three greatest discoveries of your career?" expecting him to brag proudly about something he discovered that saved the company millions. I was surprised when his response was:

"I didn't realize how really badly GM's old vehicles polluted."

"I didn't know the vehicle's running losses were so great [emissions off the engine and out of the fuel pipe]."

"I didn't realize the catalyst was so overwhelmed when stepping on the gas to enter a highway."

Probably Jon (last name withheld) has many letters and reports on which he collaborated completely denying that GM cars were major polluters, with more letters defending the vehicle emissions and never acknowledging the running loss and catalyst problems. Too many GM engineers jumped to a conclusion they knew would please management without actually testing and measuring every possible facet or phase of a problem.

## Sue First, Solve Second

Another PhD was hired by the Corporate Industrial Hygiene Staff to fight all asbestos exposure cases filed by employees. I went through the Ohio brake plant with the director of Plant Environment Staff and covered my face with my handkerchief as soon as I discovered asbestos all over my blue suit.

He said, "Don't cover your face, [the employees will] think there is a problem."

You bet there was! Not long after that, asbestos air quality laws were passed, the plant employee exposures were easily eliminated, and the labor of handling asbestos reduced with a vacuum tube transport system. The PhD staff member, hired to fight lawsuits, bragged that he won every case until 1995 when new evidence of lung disease surfaced. After that date, GM lost every case and the employees suffering exposure to asbestos were granted compensation. Proactively solving the asbestos problem would have protected the employees from developing cancer and saved the company from paying compensation to retired employees for life.

Water Pollution Assistant Director Philip Gerwert, although a member of my department, began siding with the GMAD Divisional Central Office staff who were resisting the EPA's implementation of wastewater pre-treatment standards for 13 large assembly and manufacturing plants. Upper management was too busy with high financial losses, the dropping market share, and the Paint Plan troubles to ask the right questions. More than 18 other GM plants had already met the pre-treatment standards—why had not these last 13? Three years of engineering time and attorney expenses were spent fighting and filing a legal suit over the EPA wastewater pre-treatment standards for the 13 plants. Gerwert, finally believing that the standards could not be complied with, convinced management, who sent five of the very highest in management—two VPs, two group VPs, and the vice chairman of the board—to see the new EPA administrator, Bill Ruckelshaus, and request that he look into seeking relief for the plants. Note: Ford and Chrysler were not with GM in this case.

The night that management got back to Detroit from Washington DC, the Water Pollution Control manager discovered that the GMAD Division's Central Office staff had purchased sand filter tanks and had them in storage to comply with the law in case GM's request was turned down. This meant that GM *had always been able to comply with the pre-treatment standards.* The following day, my vice president called the

EPA administrator to say that the corporation could comply with the standards but may "occasionally miss the mark." It is difficult to gauge how much influence and reputation GM lost with the EPA and its new administrator, but it was considerable and an embarrassing experience.

In the grand scheme of things, this strategy to sue the EPA and hire consultants and lawyers to resist regulatory control completely backfired because in time, the demands of the EPA lined up with the demands of the public. Think about vehicle safety and emission controls that are in use today, and the benefit they have provided to society. Now, consider GM, and only GM, winning one of those regulatory fights and not installing seatbelts, or air bags, or emission controls, or providing better fuel economy, etc. How much sooner would GM have been in bankruptcy? Where was consideration of the customer? The laws and regulations actually protected GM and the other automakers from wounding themselves sooner with the sword of self-interest.

The company has a democratic right to question any regulation, it's a good business practice—the company is the hammer, the EPA the anvil—but not with short-term profit or an even playing field for all auto manufacturers as the primary objective.

Near the 1987 Paint Plan deadline, on my arrival back from vacation, I was told to go downtown to the Motor Vehicle Manufacturer Association (MVMA) conference room right away. I was told that members from all the domestic auto companies had been waiting for me since 8:00 a.m. Why Dave Aldorfer, my employee and friend, did not call me on Sunday night to alert me to this meeting bothers me to this day.

The MVMA proposed to sue the EPA over the methods of calculating "waterborne equivalency," a pollution emission standard that had been used for the past six or seven years. The members were advocating a change in the calculation, but I knew that this change would result in a very small increase in the discharge limit. The benefit to the manufacturers was not worth the fight, but desperate people do nonsensical things.

Aldorfer made a 20-minute presentation trying to explain the calculation and the proposed change. It was my observation that none of the engineers and certainly none of the lawyers could understand, much less replicate, the calculations Aldorfer was explaining. Suddenly, Mr. Sussman of Ford Motors spoke up: "Stop! This meeting is about Nick not supporting us. Let's hear him say why."

I gathered my thoughts and replied, "Waterborne technology is improving around the world and any small advance eliminates the need for recalculating; and in technical nuances as this, the courts have always sided with the EPA." I should have said to Aldorfer, "You lost two court cases already, why try for a third?"

The Chrysler attorney spoke first saying, "We're not going to sue," and the Ford attorneys followed suit immediately. Meeting over. Finally a sensible decision was made, but not before millions were spent resisting compliance.

## Misplaced Priorities

GM still resists any pollution control measures, even those that are wise to enact, because "we want an even playing field." They insist that everyone must be forced to comply, not just GM. Unwittingly, the implication of this philosophy is that "we want a law, we want regulation; we want government in our business." This "even playing field" philosophy draws government into all parts of the business, with the result that today employees and executives turn around and complain there is too much government.

In 1984, I had the privilege of traveling with a team from GM to visit the Toyota headquarters in Japan. We met the CEO of Toyota and visited two assembly plants. One noted distinction of the Japanese approach was in the area of government regulation. Toyota's philosophy was to solve public relations problems on their own initiative before the government had time or need to get involved. For example, when a community complained about noise, an entire plant was soon coated with acoustic tile on all sides and the top. With this attitude, Toyota avoided the burden of government regulation and the expense of fighting against it.

What would have happened if those General Motors vice presidents and directors were given different marching orders? All were very bright, with superb academic credentials and PhD degrees. Imagine these individuals being told upon hire to represent the environmental and safety concerns of the public, carefully weigh the possibility of more regulatory control, and seek to have solutions developed that would give the company the edge over the competition on enactment of laws. What if they had been charged to use their brainpower and energy in a positive way, and leave the negative fighting to the lobbyists? As we will see when we fully consider the Paint Plan, the pursuit of solutions rather than lawsuits would have saved GM from a $1.5 million fine. It was the industry's first fine over $1 million. Two days later, a simple paint change was made and the plant was in compliance, negating the complaint in the suit. This legal case and all its associated expenses could have been avoided! Did no one in upper management foresee this possible outcome? Was it pride that blinded the legal staff to the possibility that the case could be lost? A prior suit against the EPA for a parallel situation went all the way to the Supreme Court and lost nine to nothing. A first-year law student reviewing the facts would have recommended that these cases not be pursued or supported. At GM the young attorney who had led the failed charge was promoted.

The legal staff had some great attorneys, but a number supported the plants no matter how they violated the law. Self-protection was the top goal, and these fellows were rewarded while other good attorneys were forced out. J.J. Hollis, a great attorney, once told Joseph Godfrey not to report fake costs to control a problem and was consequently labelled as "not a company man." Doesn't the legal staff exist to advise and recommend what is right or wrong? Or is the legal staff there to blindly defend all corporate actions without question? When GM loses in court, doesn't that hurt the company?

Staff members lived confusing corporate lives. They supported management over the raw edge of logic and the law even as the ship was sinking, resisting in order to demonstrate that they were company-minded. These are just some of the stories of staff fighting regulations, non-product issues, and anything that added cost to the vehicle, despite the fact that most safety and automotive emission regulations actually added real value desired by the customer. Supposedly, management gets the big bucks for making the toughest decisions, but when staffs submit evidence of agreement with management's predisposition on the subject, there are no hard choices. There are no strong presentations to management of facts supporting the proposal of safety and emission controls, no vigorous debates between executives—only "team players" making management jobs really easy.

The environmental staff should have presented the best side of why regulations were being proposed in Congress. The industrial relations and industrial hygiene staff should have fought for the control of oil mist, asbestos, and noise pollution in order to protect the company's employees and eliminate compensation claims. The legal staff should have advised GMAD to comply with regulations rather than waste time and money in a fight. When staff are given the freedom and encouragement to do the job to which they were assigned, the entire company benefits, not just a plant or an individual in management. The self-protective culture as it manifested itself in the corporate leadership did not encourage consideration of the employee, the customer, or the company as a whole. While other corporations rally around a core value such as innovation, customer satisfaction, or product quality, all of which have been proven to reap profits over both the short- and long-term, the big three automakers were united in their fierce resistance to regulations from the government. Regulation was the common enemy and at GM, I observed resistance to be the unifying goal.

Just like they fought against critique by the EPA, GM leadership resisted critique from fellow employees and fought against critique from the customer. The culture of corporate pride blocked any negative feedback or questioning of the status quo. I have experienced how engineers and executives are badly treated for proposing that the company cooperate with the governmental agencies on emission standards. The corporate culture made it clear to everyone that the rewards go to the fellows who are resisting, not assisting, the agencies. Management's resistance to critique was a factor in GM's demise in general but most particularly in the unquestioned approval of the outrageously costly Paint Plan.

# 8

# GM's Disregard for Consumers

Just as the culture at GM fostered the belief that management knew better than the board, and GM knew better than the government, this same pride manifested itself in an attitude that it was acceptable to deceive the public. There was a prideful assumption that the customer would always be there and this led to a disrespect of the customer.

## Deception Begins

The first telltale event involved upper management and most of their supporting staff in a decision that changed the way the auto industry sells vehicles. The fallout from that decision should have provoked an evaluation of board oversight and management decision-making. The story involves the Olds Rocket 88 engine case. In 1959, an Oldsmobile racecar won the first Daytona 500. GM used this event to advertise the Olds Rocket 88 to great sales advantage. The Rocket 88 became a household name and the Olds Division sales surged to fourth place in the U.S. market. Sales were so good in the '70s that Olds needed more engines. But increasing production of the 88 was too costly, so a suggestion was made to use a Chevrolet V8 that was in sufficient supply. Up until this time all car engines were made by that particular car division: all Cadillacs had Cadillac engines, Pontiacs, Pontiac-produced engines, and so on. Installing a Chevrolet engine in an Olds 88 vehicle was a serious matter. I owned a used Olds 88 and in the course of discussing it with Bob Rogers, executive engineer of Automotive Safety, he shared with me the following details of the discussion that took place in the Chevrolet engine replacement meeting.

Q: Why not announce to the world that the Chevrolet V8 was a great engine, as good as that of the 88, and not hide the fact about the switch?

A: If we do, Olds will lose sales!

Q: Can we make sure loyal customers of the Olds 88 Rocket engine get an 88?

A: No, the word will get out to everyone!

Q: Should we notify the dealers both Olds and Chevrolet?

A: No, they'll work it out.

Q: Legal staff, is there a risk in not telling the customer?

A: No, not much!

Was there not one company man who foresaw what would happen when he had to get his personal vehicle fixed at a dealership; not one engineer who knew that dealers stocked different parts for their particular engine; not one fellow to say it is wrong to lie to customers; not one soul brave enough to take on the "short-term" sales department mentality and make some waves? No, not one!

In a short time, Olds buyers with Chevrolet engines needing maintenance or repair were told by their dealer to take the vehicle to a Chevrolet dealer who had the proper parts. As a result, a class action suit and many individual suits were filed. GM responded by hiring 300 lawyers (including the first woman in GM's staff). The court decision required GM to issue coupons worth hundreds of dollars to most of the complainants with owners suing for more. Despite a final cost of well over $30 million and a loss in reputation, the end result was not all bad. The coupons applied to the purchase of new GM vehicles, resulting in a temporary increase in sales, and positive changes were made in the way parts were manufactured and tracked. For hiding the truth from consumers, the judge ordered a wholesale change in the way the industry was to inform the public. GM had to take out full-page ads in all the major newspapers for weeks telling the world what they had done. The secret was out! All dealers had to display a large plaque in their showroom windows announcing what happened and what to expect from that point on. The greatest resulting innovation was that every vehicle had to have a window sticker itemizing all the important features, including where the car was made.

Management has available to them a number of staff to advise them on critical decisions. The corporation has financial and legal staff as well as those specializing in Industrial-Governmental Relations, Public Relations, and Industrial-Labor Relations. All are available to analyze issues and make recommendations. But as exposed in the Rocket 88 case, the staff supported management's predisposition to make money and sales in the short-term rather than think hard about the long-term. Could a comparison between the Olds 88 versus the Chevrolet V8 have been used to increase the sales of both engines? Probably. This episode should have caused some of the board members to question how issues are debated and presented by staff, particularly when they involved major product programs. But I doubt anything has changed. The board remains reactive rather than proactive. A scan of the GM headlines in the last twenty years indicates that the only time the board of directors gets involved is after the fact.

## Deception Allowed

To my knowledge, a culture of deception has never been challenged by the board, hence the problem of deception did not end with the current customer. GM attempted to hide the truth from the general public, never considering that the general public is made up of current and potential customers. As I write these memoirs, plant problems have been in the newspapers for months, and management has received letters from the citizens threatening lawsuits. In comparison, problems associated with paint systems seem almost humorous today, like the community odor problems at the new Pontiac Truck plant built in the '60s.

A new assembly plant was to be built directly across the street from a well-to-do new neighborhood in Pontiac, Michigan. The residents were concerned about the noise and traffic as well as the odors from the future facility. Argonaut, GM's real estate division, prepared a rendering—a beautiful picture of the future plant. There was not one discharge stack on the roof—it looked great! But it turned out to have many, many stacks producing lots of odors.

The problems got into the newspapers, so a meeting was called to do something about them. It was held in the executive "Fourteenth

Floor" boardroom with John DeLorean, Truck Group vice president, and Alex Mair, regional vice president. With me, were engineers from the manufacturing staff assigned to develop environmental controls. The plant manager started out by saying that I wanted engineers to go out into the neighborhood, take samples of the odors, and test the plant's stacks for odors. He denounced my plan with an articulation that "We don't want to admit the plant is the problem."

DeLorean was lying back in his chair making a paper airplane out of an 8.5 x 11 sheet of paper and said, "Are you engineers? Go out and take samples and fingerprint them. Match them to the stack and control it."

Mair said, "The plant problems have been in the newspapers for months, management has received letters from citizens that we may be sued. I think they know we have a problem! By getting out in the neighborhood and taking readings, we will demonstrate that we are doing something."

I thought, *Why are meetings like this necessary?*

If directed by management, the Industrial Relations staff could have enhanced GM's image and reputation worldwide by ensuring that GM's local tax dollars were being spent in a manner that represented the corporation's good name to the community at large.

## Communities Not Cared For

General Motors was the major source of jobs and income in three major cities: Detroit, Flint, and Saginaw. Flint especially was all GM. In 1993, the FBI published a list of the most crime-ridden cities in America and sadly Saginaw sat at number one, with Flint and Detroit not far behind.

In the 1960s, I heard of a meeting between a GM representative and the Saginaw tax assessor, with a request to lower the assessment of the Chevrolet foundries, but the county assessor stated the plants were well below any reasonable tax figure, so it couldn't go any lower.

The GM fellow said, "Oh! Let's go to dinner."

The company culture was displayed in many ways to the general public, and no one in management seemed to make the connection to vehicle sales.

## Warranties Not Honored

Forty years later, this attitude persisted. In 2008, a *Detroit Free Press* headline read, "GM Won Its Case for the Fourth Time." Four states had filed class action suits over leaking head gaskets on many 6-cylinder GM engines. The problem with the gaskets started after the company changed the radiator coolant from the normal green type to a new orange color coolant called DEX. A headline read, "GM Beats its Customers." Some great business advertisement! It was almost as if the customer was an unavoidable nuisance.

A short time later, a fifth state, California, sued GM in a class action suit, resulting in the court deciding all verified complaints must be reimbursed for head gasket replacements. My family had three vehicles with head gasket problems. One of the vehicles had to have the head gasket replaced twice. The second repair was different from the first, in that the gasket material and bolt tightening specification were improved. GM engineers finally found the solution, but not before millions of cars were affected. I had applied for $2,400 to $3,000 in compensation for my family's repairs, but received nothing because of the bankruptcy. I do not believe that the money GM lost in making the repairs came close to the loss they suffered in reputation and sales.

Another illustration of how GM tended to choose short-term profit over long-term customer care was how they dealt with and honored their warranties. My power steering assist mechanism failed on my 2004 Malibu, and I was asked to pay $900 to replace the entire steering column. GM had changed from a hydraulic assist to an electric-powered unit, and failures occurred on early models. My vehicle VIN number fell just out of the range of vehicles recalled for a free new column, so it was not covered, even though the problem was exactly the same as on the other vehicles. The regional sales office was contacted, and my payment was reduced to $300 but it still left a bad taste in my mouth.

## Customers Devalued

For several decades, within GM, as these examples have shown, no one was upholding the value of quality workmanship. No one was valuing the customer. Everyone sought short-term profitability over long-term customer loyalty. Historically, cocooned within the American-made auto bubble of Detroit, GM's pride blinded it to the truth that outside of Michigan, customers were abandoning generational loyalties in favor of the quality of the Japanese-made product. In recent years the quality of GM's vehicles has greatly improved, but warranty problems still occur and GM's response has and will reflect its true character.

Executives in high positions could have redirected the company at any point. In their day, Roger Smith and Joseph Godfrey only contributed to the problem. They were pushing for a "wow factor" without consideration of the nuts and bolts. What would be genuinely valued by the customer? What would be the actual investment costs of their grand schemes?

To appreciate the ludicrous nature of the Paint Plan it is necessary to have some knowledge of auto production planning, automobile paints, and the operation and regulations of the EPA. It would appear that the selling point of the Plan was that some form of paint change was being required by the EPA and a complete modernization would put GM ahead of Ford and Chrysler, thereby forcing them to play catch up. The goal was to have GM known as the world class producer of the most glamorous, high quality, low cost vehicles in the world— and there were no restraints on the funds for the plants and no reality check to see if the Paint Plan would bring about the desired result. It seemed to be all about reputation and not actual profitability.

# 9

# Automotive Painting

## Normal Product Planning

Designing vehicles, manufacturing, assembling, and ultimately selling vehicles at the highest profit margin is the primary goal of any auto company. Therefore, the sales forecast for the vehicle offerings should drive all phases of the company's operations. Employee hiring, product development, materials procurement, selection of process systems, and, most importantly, the scheduling of capital expenditures, all should take into consideration the short-term and long-term desires of the consumers. You must produce a product that the customer wants to buy, but from conversations around the lunch table, the customer was not the primary concern of my peers in GM management. Our concern was Ford and Chrysler. We hated Ford and Chrysler. Nonetheless, certain production details were dictated by the nature of the product.

Years ahead of the mass production of a new vehicle, a product plan is set and the powertrain (the engine and transmission power system to the wheels) must be designed and manufactured in accordance with the schedule established in the product plan. The assembly plant to produce the finished vehicle must be selected. A sound fiscal product plan should include strategies to optimize the use of existing equipment, to minimize the retraining of the workforce, to reduce the cost of shipping both of materials and vehicles in and out of the plant, and to make use of the plants' downtime to upgrade the technology of the manufacturing systems. GM in the 1980s was generally unconcerned about optimization. It wasn't until November 2006 that I read in the *New York Times* an article reporting the travels of GM executives to China to "learn the secrets of its low-cost production."[15] With its vast storehouse of experience, shouldn't GM have been the expert?

## The Paint Process Line

Henry Ford hand brushed the paint on his vehicles. Today the process is fully automated. In the first step of the modern automotive paint process, the naked sheet metal body is washed with a strong detergent, which removes grease and oil, rinsed, and then washed with a mild acid to clean the pores of the metal. It is rinsed several more times to remove all traces of solids or salts. After the final rinse, it is quickly heat dried to preempt any rust formation, and then a microcrystalline zinc-phosphate coating is applied to reduce any future rusting and improve the adhesion of the paint layers. The final coats provide little protection from rust damage without this pre-painting process. A primer is then applied and heated to set the chemical process and to create the optimum surface for a smooth and lasting topcoat. As stated in a 1978 GM study from my files, "The initial primer application is critical to the performance of the total paint finish. Until a few years ago, the basic body shell was painted by hand spray operations, and other components were primed by dipping and flow coating processes."[16] In flow coating, the parts were passed through a tunnel with dozens of nozzles spraying out paint. The parts were thoroughly coated but so were the conveyors, hooks, and tunnel walls. A waste pan caught even more.

The first major industry improvement was dipping of the sheet metal: body, panels, or parts, into a tank of paint. To improve the efficiency, an electrical charge was given to the part and the opposite charge to the paints. This process, which is known as electrodeposition, is referred to as ELPO in the industry. It increased paint application efficiencies to 98% for these foundational coats. "Following the successful development of the electrodeposition process in the late 1960s, GM has been converting the primer process to electrodeposition as facility modernizations have occurred."[17]

The coat that you see is the color coat. GM offered customers approximately 15 colors for truck lines and roughly 16 for each line of passenger cars. Historically it was called the topcoat and is

FIGURE III-1

TYPICAL PAINT SHOP OPERATIONS

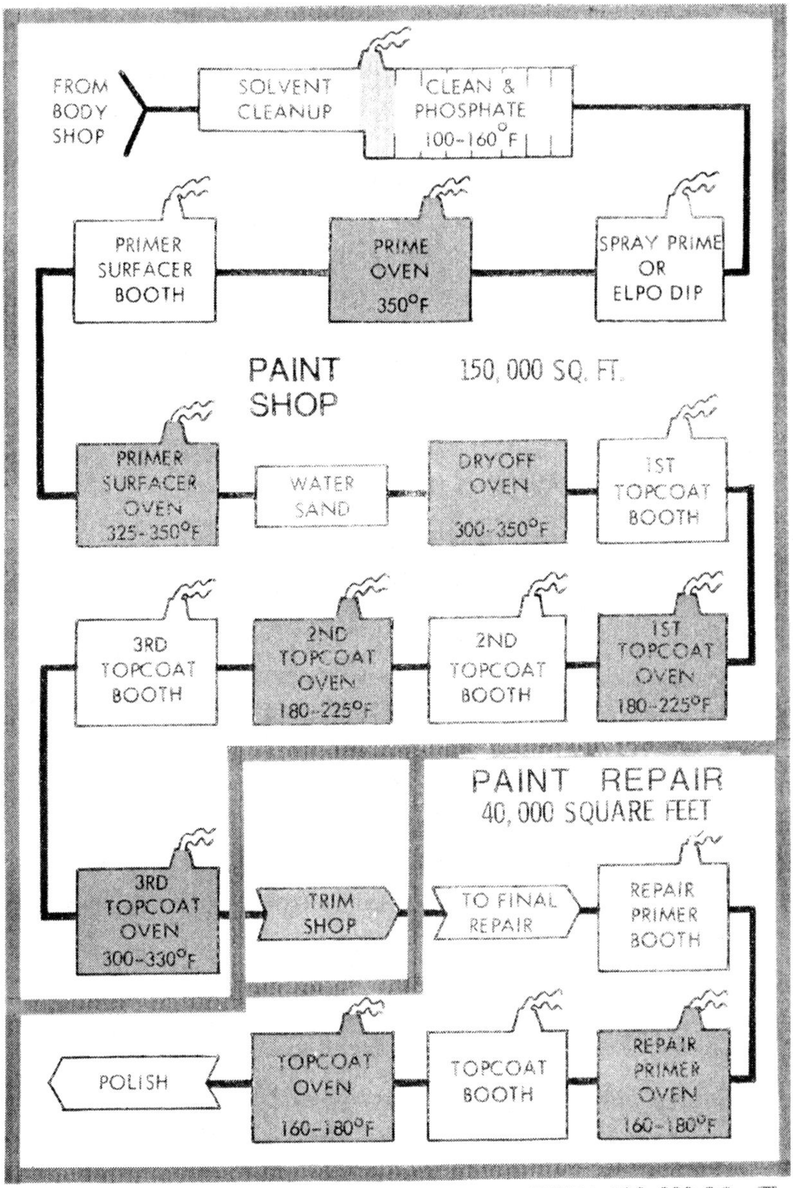


TOTAL 190,000 SQ. FT.

Credit: GM Internal Study, 1978

*Part of an internal report on automotive paint and the EPA—this sketch shows the steps in the automotive paint process at the beginning of the Paint Plan.*

commonly referred to in that way, but in its modern form, you will see it referred to as the BC/CC or the base coat/clear coat. This two-layer coat was introduced by the Japanese in the early 1980s, a few years after the date GM had promised it would settle on a new form of paint that would meet the air pollution standards. The base component of this combination is the coat that gives each car its color. To increase the hardness of the final surface and add some glamour, a "clear coat" of powder paint is applied over the base coat. Although it is only a couple of thousandths of an inch thick, it is the colored base coat that is the most expensive, and the greatest source of air pollution issues.

As each paint layer is dried, the oven exhaust stacks emit some degree of odor. When the odor is strong and offensive and the neighbors complain, the exhaust must be run through an incinerator or through chemical scrubbers. But not all odors are pollutants. The only major hydrocarbon pollution source is the colored base coat. For a paint shop to reduce the VOCs (volatile organic compounds) emitted and meet the regulatory limit, there are basically three options: 1) change the paint formula, 2) improve the efficiency of the paint application, or 3) add an abatement control device to remove the emissions from the exhaust before final release into the atmosphere. Of these three options, GM placed most of its hope in changing the paint formula and secondarily in improving spray efficiencies. But the first objective the paint superintendent *should* have is to provide every vehicle with a world-class finish that will help sell the product.

## Topcoat Paints

In order for any paint to be applied smoothly and evenly, the pigment must be diluted or thinned with either a hydrocarbon solvent (think of oil-based paints) or water. In auto manufacturing this diluting solution is then drawn out of the paint in an oven, leaving only pigments and base materials. There were two main categories of topcoat paint used in the automotive industry in the late '70s: lacquer and enamel. While the end result may seem identical to the untrained eye, the source and chemical structure of the two paint types are very different. "Lacquers are thermoplastic materials in which the film-forming resins are dissolved or dispersed in solvents. Enamels are also dissolved or dispersed in solvents, but are thermosetting materials which require heat-curing to complete this film."[18]

Within the varieties of chemical compounds that can comprise a paint, the two main components are the solid component containing the pigment and hardening compound, and the solvent. It is the solvent that is the source of the hydrocarbon pollutants (VOCs). The ratio of solids to solvents is inversely proportional. Looking at the graph (page 94), one can see that this ratio could be expressed on a continuum going from GM's original lacquer paint which was only 13% solids (pigments and binders) with 87% solvent discharged in the drying process to the outside air with 10–20% of the total paint sprayed going into sludge waste. At the other extreme are powder paints that are applied to most home appliances and approach 0% solvent. Considering these same extremes, the lacquer paint that is 87% solvent is known to release 52 pounds of organic hydrocarbon for every gallon of solids, whereas powder coatings release essentially zero.

## LACQUER PAINTS

Lacquer paints were first developed for automotive use in the 1920s and continued to be popular for the next 30 years. Even when dried at room temperatures, lacquer forms a hard durable finish that can be polished to a high gloss. This was undoubtedly the paint of choice for Henry Ford since his cars were hand brushed but had a reputation for shine. In automotive use, the lacquer polymers are dissolved in VOCs, such as nitrocellulose. Lacquer paint can have varying percentages of solids from 13% to 27%, but all are thinned with a hydrocarbon solvent and this solvent is released into the air during the drying process. To those that restore or repair auto bodies, lacquer carries a significant advantage. The fact that enamel paints are "thermosetting" means that they change chemically and irreversibly in the heating process. In contrast, lacquer paint can always be reheated causing the paint to reflow to a glamorous finish. This reflow characteristic of lacquer paints historically gave GM passenger vehicles a competitive edge over the cars and trucks that were manufactured by other auto companies and finished with enamel thermosetting paint.

If a GM car had a dent or scratch on the door panel, the affected area could be sanded, repaired, and repainted with more lacquer paint. The area containing the new paint could then be reheated and the paint would flow to cover the scratch and blend in to the surrounding area. It could be repaired relatively quickly and easily. A vehicle with an enamel topcoat and the same damage must have the entire door

---

*This graph illustrates the varying amounts of pollution created by the available paint formulas that were available in the '80s.*

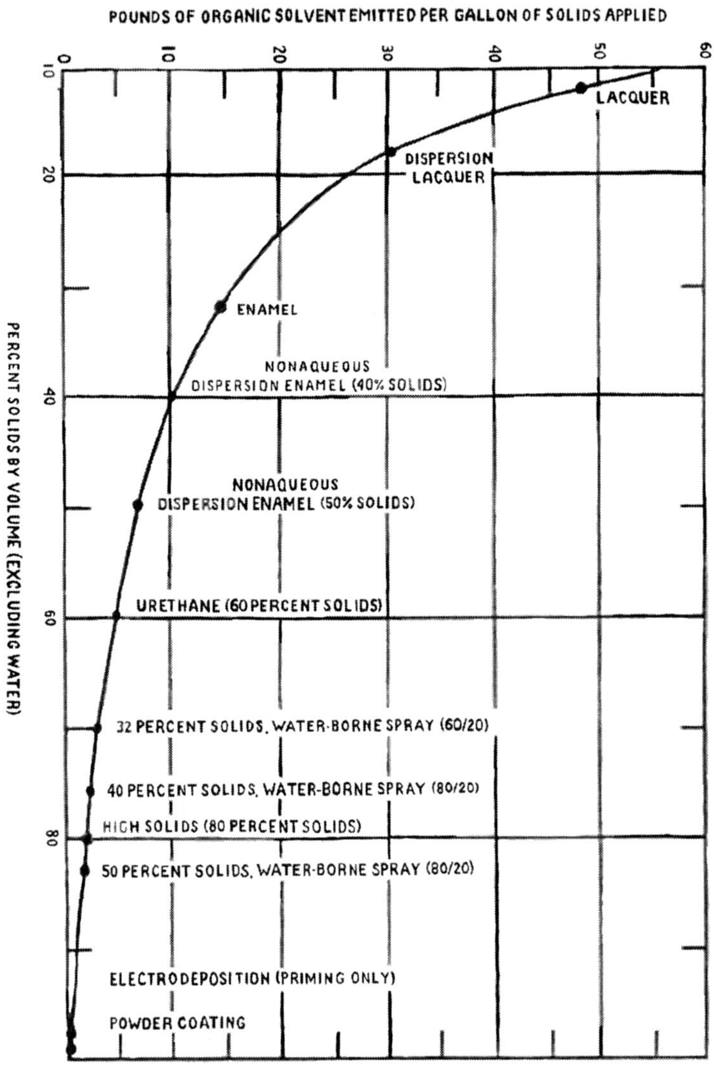


Figure 1.2. Percent of solids versus pounds of organic solvent emitted per gallon of solvent applied. (Assumption: solvent density = 6.6 lb/gal).

Credit: GM Internal Graph

panel re-sanded, the repair made, and then the entire panel re-painted and re-finished. It is a much more laborious and expensive process.

Yet while their paint surfaces were more difficult to repair, GM's competitors did not have the issues that GM did with the paint emission standards of the EPA. Their enamel paints were not emitting the volume of VOCs that were coming from the application of lacquer paint. According to an internal GM memo dated April 24, 1978, "Enamel paints used by Ford, Chrysler, and American Motors emit only one third of the hydrocarbons emitted with the use of lacquer paints" ("Report on General Motors Corporate Paint Plan," 4/24/78). Without looming compliance deadlines these competitors were free to schedule any upgrades to their paint shops in conjunction with plant closure cycles that matched their overall product or business plan. GM, it seemed, did not have this luxury.

## Enamel Paints

It was concern about the emission of VOCs in the paint drying process, along with encouragement from the paint suppliers, that eventually drove the entire industry to move to enamel paints. Their basic chemical makeup caused them to be less polluting. Like lacquer, enamel paints are also appropriate for hard surfaces that require a durable high gloss finish. They can have an oil base or a water base solvent. Enamel paints ranged in solids content from 32% up to 98% for powder paints. Ford, Chrysler, and all other auto companies including the GM truck division applied enamel coating with a solids content of 32% and solvent content of 68%. This ratio results in a 65% reduction of polluting hydrocarbons over GM's passenger car lacquer coatings. This lower emissions level was within the range acceptable to the EPA for all existing paint emission sources in GM.

A variety of enamel paints were developed in an effort to reduce the percentage of solvent. First solvent-based enamel coating with as high as 40% solids and 60% solvent became available. Then with the addition of water-based enamel, the percentage of solvent dropped to 30%, then 25%, and finally 20%. At the same time, the percentage of solids increased.

Keep in mind though that each increase in solids might require a change in the sprayer or applicator, in the speed of the production line, the temperature and time in the ovens and, most critically, a change in the appearance of the finished surface. To some at GM, including Godfrey, all of the enamels seemed to have "an orange peel" surface with a series of minuscule craters. But at the dealerships of our competitors, when customers had concerns that the appearance didn't match glossy smoothness of the lacquer, dealers would point out that the new paints were thicker and provided a harder, more protective finish. To this there was some truth.

## Powder Paints

The best ratio of solids to solvent would be achieved with the development of dry powder coatings at 98% solids and only 2% solvent. For all the paints that are applied as a liquid (lacquer, enamel, and waterborne paint systems), there is overspray that is captured in a waterfall collection bath. This results in polluted water and the generation of paint sludge waste. Also, as discussed, these systems all produce air pollution and community odor problems. The environmental advantage of powder coating is that it is a dry spray system. Dry paint over-spray can be easily captured. Resultant air and water pollution is essentially reduced to zero.

The Europeans were the first to pursue powder paint as a solution to the pollution issues of liquid paints, and a German scientist garnered the first patent in 1955 for a powder coating that was applied via a fluidized bed.[19] As early as the late 1960s, I had GM Manufacturing Staff, Paint Research and Development manager Gene Leithauser working on powder coatings, but Joe Godfrey had this department placed under his control in 1974 and the research stopped. Imagine if the development of powder paints for autos had been pursued, so many problems due to the Paint Plan could have been avoided. GM could have been on the cutting edge of environmental concern. Now that would have been "leapfrogging" the industry!

Interestingly, powder paints are now used on 95% of all painted products that you find in stores: refrigerators, stoves, washers, small appliances, etc.; but they were not seriously considered for application to vehicles until the early to mid-1980s. Following several years of research and development, the Sames Co. of France, the first manufacturer of a spray gun for application of powder paints, was enlisted by the Japanese to apply spray powder paint to vehicle bodies in 1993; and the Swedish truck manufacturer, Scania, implemented a powder coating system in 1995, a mere 10 years after the Paint Plan era.

When the industry followed the Japanese in switching to a BC/CC (base coat/clear coat) finish, they all took advantage of the fact that the clear coat could be applied as a powder coat. The high level of protection provided by this final clear coat allowed the base coat layer to be very thin. The thinner layer equals less paint and less paint equals less pollution.

# 10

# POLLUTION AND THE EPA

## UNDERSTANDING AIR POLLUTION

IN THE FALL OF 1957, SIX MONTHS INTO MY CAREER WITH GENERAL Motors, Fred Russel, my former employer, invited my wife and me to visit Weiser Lock Company in Watts, California, with an offer to become his industrial engineer. I vividly recall that the smog pollution was so bad that it was impossible to stay outside the hotel for more than a few minutes before our eyes and throat began to burn. With no regrets, I decided to stay with GM and keep Michigan as my home, but the trip increased my understanding of why pollution sources had to be dealt with.

The Los Angeles smog, like the smog around many other world cities, is largely composed of fine particulate and a chemical mixture of hydrocarbons that produces ozone when radiated by the energy from the sun (photochemical smog). In the *upper* atmosphere, ozone is beneficial because it filters out some of the UV radiation from the sun, but in the lower atmosphere, it is harmful to both plant and animal life. Ozone, a chemical structure consisting of three oxygen molecules ($O_3$), can cause coughing, choking, severe headaches, and eventual death due to lung disease. It has also been shown to reduce crop production. Facing unique topographical conditions, and a plethora of vehicular traffic, California was the first state to force the use of water-borne paints in order to reduce the VOC emissions from industrial paint facilities.

Vehicle emissions were the largest single pollution source accounting for over 50% of the total emissions in the L.A. basin, but painting ran second. According to an article published in the *Journal of*

*American Oil Chemists' Society*: "The Air Pollution Control District of the Los Angeles County inventoried the organic solvents being emitted into the atmosphere in a 24-hour period in their county."[20] The study determined that "the major [non-vehicular] contributor to the 550 ton daily emission of organic solvents was industrial, commercial, and residential painting."[21] As a result of the data from the study, "California Rule 66" was codified on July 28, 1966, and the two California General Motors assembly plants—one in Van Nuys, California, and the other called "South Gate" in Watts, found themselves under stricter emission standards than any other GM plants in the U.S.

But by the mid-1970s, Los Angeles was not the only city with pollution concerns. Chicago, Detroit, Pittsburgh, and New York were also experiencing days with high ozone levels. The result of these widening observations was an establishment of acceptable standards for all emitters of hydrocarbons. For example it was determined that GM's Flint, Michigan, plant emitted 59 tons per day of hydrocarbons—the equivalent of the emissions of 4.8 million cars (with the EPA mandated catalytic converters).

In June of 1976 the EPA's Office of Air Quality Planning and Standards (OAQPS), under Don Goodwin, began work on Control Technology Guidelines (CTGs) for vehicles, assembly plant paint systems, oil processing facilities, etc. According to a friend in that EPA department, John Calcagni, David Hawkins, Asst. Administrator for Air and Waste Management, took the memos prepared by Goodwin and assigned strict numerical regulatory limits to each industry category. The OAQPS issued these standards in the summer of 1977. It was then up to the regulatory agency of each state to incorporate these guidelines into an SIP or State Implementation Plan.

The states were to use the guidelines from the EPA but had the latitude to work with the industries in their state to determine what constituted "RACT" (Reasonable Available Control Technology) for an industry's existing pollution sources. In a document titled "The Early History of the CTGs" Calcagni writes: "In most cases involving coatings these limits were based on an 81% reduction of the VOC in typical coatings in use. The 81% reduction was based on 90% capture and 90% control since abatement was always an alternative. For automotive assembly plants the number was based on a GM facility in Fremont, California, using waterborne coatings. State VOC regulations...were due in SIPs by January 1, 1979, and compliance was to be as expeditious as practicable, but not later than December 31, 1982." When these guidelines became law, I knew that GM would be forced to change its paint and application systems or reduce its paint emissions with "stack" (as in smoke stack) abatement devices. This second option, however, had been eliminated by Godfrey. More about this later.

As a side note: a typical GM passenger assembly plant in the late '70s emitted 6,000 to 7,000 tons of hydrocarbons per year, compared to a 2,500 to 3,000 total hydrocarbon emissions from a similar Ford assembly plant. In 1987, the Chrysler Jefferson Avenue plant was limited by the state agency to emit only 600 tons a year. In 2007, the BMW plant built in South Carolina was limited to 274 tons of emissions a year. Real progress has been made in the industry since the early days of regulation.

## EPA Terms and EPA Demands

The Clean Air Act was originally put into effect in 1963, but the amendments of 1970 had the greatest impact on the auto industry. These amendments mandated comprehensive federal and state regulations in regard to the emission of air pollutants while granting increased enforcement authority to the EPA and its state level agencies. The EPA worked in coordination with each state in order to measure the current pollution levels and then develop strategies to reduce the pollution. Cities around the country were analyzed to determine their ozone levels. Each locality was designated as having "attained" (or always been within) an acceptable ozone level, or as a "non-attainment" area where immediate action would be required in order to make the ambient air less harmful to the residents. The standards imposed on local industries were different in attainment and non-attainment areas. In the priority "non-attainment" areas, existing industries were told that they must meet a certain numeric emission standard that was determined to be within reach. All of our GM facilities were in non-attainment areas with the exception of a plant near Atlanta.

The first level of control that applied to stationary pollution sources already in operation was, and is, the level deemed to be reasonably attainable and therefore called RACT. Operations were granted a five-year period in which to meet the current RACT standard. In theory this was a concession to industry to set compliance with the standards at an attainable level that could be reached without burdensome expense and within a reasonable time frame. Despite the outcry from industry, the EPA did recognize the economic importance of U.S. manufacturing plants and the intention was to work with them in a reasonable manner that would benefit the public. But it was also understood that as technology advanced and became readily available, the acceptable emission rate for RACT could be lowered.

**CASE I:** All existing pollution sources in areas of "non-attainment" required RACT control. Since a large portion of the U.S. population was exposed to unhealthy air quality that was causing multiple health issues including some deaths, the Act required all industry pollution sources to either install abatement equipment, which would capture pollutants before they reached the atmosphere, or change their paint processes within five years. So there would be no arguments or legal battles with industry, the regulations only called for the use of controls already in wide use, easily purchased, at a reasonable cost and generally installed without much disruption to plant operations. The emission reductions experienced were 50% to 70% with some sources reduced by 90% or more.

**CASE II:** When building a new plant, modifying a major source, or adding a major pollution source in a polluted city, companies were required to do the very best to limit emissions and not to exacerbate an already bad situation. For very common industrial sources like a powerhouse boiler, auto paint system, large incinerator, etc., the EPA used data on existing facilities to set a numerical standard of acceptable emissions and used that standard to establish the New Source Performance Standards (NSPS) that could be applied to any new facility. For the automotive paint shops, the NSPS level controls limited emissions to 2% to 10% of the potential amounts the source could emit. Although the companies resisted the higher restrictions on new construction, building the best technology into the project at the time of construction can actually benefit the company by reducing the expense to a fraction of what would be needed to retrofit or make later modifications.

In summary, the EPA regulations had two major provisions: 1) All existing assembly plants' paint systems must be controlled within five years using RACT, and 2) any new or modernized system must meet the New Source Performance Standard (NSPS). The current value attached to the NSPS had been established in 1974 at the GM California plants where they were using waterborne enamel paints for the first time in the industry. But it was the first of these provisions that required the immediate attention of all the auto manufacturers.

How did the new regulations affect Ford and Chrysler? Ford Motor Company and the Chrysler Corporation were compelled to meet the same EPA paint standard as GM and were just as active in fighting EPA controls as GM was; but in the 1970s they were both already applying enamel topcoat paint to all of their vehicles, and therefore their emission levels were low enough to meet the RACT standard. They were free to schedule paint shop modifications in concert with their product business plan.

Only when building a new paint shop did the water-borne topcoat standard apply to Ford and Chrysler. This allowed these companies the freedom to install, operate, and evaluate any new paint system before another was built. They had time to make improvements to the application process and to the quality of the result, and history will show that this is what they did.

When the new Act went into effect, the GM truck plants were applying enamel paint and were already in compliance, as was the California plant. But all of the other car assembly plants were using a traditional lacquer with an emission rate that exceeded the new standards. Clearly, GM would have to act in some way to bring the plants into compliance or else potentially face millions of dollars in fines. The path chosen by GM was to request 10 years to develop a *completely new paint* that would be very low in emissions. In conjunction with the development of new paint, the plan was to upgrade all GM paint shops to the latest technology. It was an ambitious plan that appeared forward thinking on paper but was never carefully evaluated for its costs and feasibility. Consideration of other options was never discussed by the Emission Review Board (ERB) or any other body with authority. But there were other available options and I tried to bring these to the attention of management.

**Option One: Solvent Based Enamel**

The first option for GM was to agree to convert the lacquer facilities to the enamel that was already in use in the Fremont plant and was known to meet the RACT standard. But to implement this plan GM would need to negotiate with the EPA for 10 years instead of 5 to allow time for the physical conversion of the paint shops and their spray systems. The painting systems would have to be changed to accommodate the chemical differences between lacquer and enamel paints, but the needed technology and paint was already in use within GM. GM truck plants (nine in number) would be in compliance immediately, leaving only the car assembly plants and some body plants in the plan. This option was rejected for being unimaginative. GM would be simply following in the footsteps of Ford and Chrysler. There was a reluctance also to give up the shine and repair advantages of lacquer.

**Option Two: High Solids Lacquer Paint**

To maintain the GM paint quality advantage there were two ways that GM could have continued to use lacquer paints and thereby avoided complete overhauls of the paint systems. The first would have been to request a separate RACT standard for lacquer paint. Because of the size of the corporation and its impact on the U.S. economy, GM could have easily convinced the EPA to grant it a separate RACT level to meet. GM was the only company in the world applying lacquer topcoat paint, but they had as many plants as the other two large U.S. companies combined. The argument could have been made that a conversion to enamel paint would be at a tremendous cost and not a reasonable requirement. Under the Reagan administration it was highly likely that the EPA would have allowed this.

Had William Chapman, director of the Washington DC Office of Industry-Government Relations been requested to seek a reasonable standard, knowing the steel industry was granted 10 years to apply RACT, the reply would have been, *What else do you want?* GM's IGR staff and Michigan's Congressional representatives were extremely effective in getting special things for the company. For instance, the company bought the Hummer, a vanity vehicle modeled after a military armored car with the gas mileage of a tank, and was able to sell a few because the vehicle was given an exemption from the gas-guzzler tax and companies could write off the cost of these vehicles from their income tax. There was no national benefit from the exemption. Now that's a company with real connections.

GM's influence was exactly what worried EPA administrator David Hawkins. From the outset he believed that GM would use the Plan to buy time like other industries—steel, utilities, and oil refiners—until Reagan was elected in 1980, and then seek complete relief from regulations under the Republican leader.

If GM had taken steps to improve spray efficiency, allow abatement odor controls on paint drying ovens, and install the new ELPO primer waterborne paint, they could have promised the EPA a 75% reduction in lacquer paint emissions with a potential reduction approaching 80%. I stand confident today that it is a regulatory standard for which I could have gotten approval, even without the aid of GM's IGR staff.

A second option would have been to request permission to develop a high-solids lacquer paint that would meet the industry-wide RACT standard by approaching the enamel emission rate. The painting systems would not need to be replaced—only modified—to accommodate the lower percentage of solvent. If the standard was easily obtainable, GM could accept the five-year deadline. If compliance proved too difficult to obtain, they could request an extension to 10 years. This option was rejected as being unfeasible, but I have in my files a letter dated September 29, 1978, from F.F. Rhue on PPG Industries stationary assuring Gene Leithauser that "PPG will be in a position to supply to General Motors either 17% or 29% lacquer dispersion coatings by the 1980 model change." Eleven years later 27% dispersion lacquer paint *did* satisfy the EPA requirements. It was developed, tested and used in 1989 to bring the Lordstown plant into compliance. Despite internal claims to the contrary, the paint proved to *reduce* the paint cost per vehicle. Amazingly, in spite of the cost and regulatory advantage, it was not utilized at Lordstown until the Monday *after* GM was ordered to pay a $1.5 million dollar fine for non-compliance at the plant. Had it been brought online sooner, the monetary fine and all the associated legal fees would have been avoided; but as the record will show, GM would rather fight than comply.

**Option Three: Use Abatement**

A third option was to continue the use of lacquer paint already in use and make no changes to the spray systems, but meet the enamel RACT standard established by the other companies by applying a minimal amount of pollution abatement control on the exhaust stacks from the paint spray stations and ovens. This option was never on the table because of Godfrey's adamant rejection of abatement: "I don't want to see a damn chemical plant on my roof!" It would have been prudent for him to consider it more closely. Others in GM simply told the EPA that abatement would be too costly. They had not done their homework and they certainly had not weighed the cost of abatement against the cost of the alternatives.

## GM's Aversion to Abatement Controls

A variety of industries have demonstrated that the EPA standard for VOC emissions can be met using abatement. The most common abatement controls in use today are 1) thermal incineration with heat recovery, 2) catalytic incineration with heat recovery, and 3) fixed or fluidized-bed activated carbon adsorption units with heat recovery. A scientific study of the third option, carbon adsorption, had been performed at the Fremont plant as part of a coordinated effort to meet the strict standards of California Rule 66, but Joseph Godfrey had ordered the study results to be destroyed as soon as possible.

In contrast, when faced with the reality that their manufacturing plants were emitting pollutants, the highly successful 3M Corporation made an executive decision to address the emissions of their manufacturing process by capturing and removing the hydrocarbons through abatement devices on the exhaust stacks. They made no changes to the formula for their existing tape products. Once the abatement devices were operating at their 50 plus facilities, they had no trouble meeting EPA regulations. With no time constraints, they proceeded to focus their business energies on developing water-based products that matched customer demand

*and* produced little or no hydrocarbon pollution in their production. At General Motors stack abatement could have solved all of the paint emission issues while allowing the Paint Superintendent to focus on quality—quality that would be superior to our competition and, more importantly, quality that would please the customer and sell more cars. If Joseph Godfrey had not forbidden exhaust stack abatement, the entire expense of the Paint Plan and all its associated legal expenses could have been avoided.

After GM entered an agreement with Toyota in 1983 to jointly build a small car for GM in a former GM plant in Fremont, California, a group of GMAD engineers and I were sent to Toyota headquarters in Japan to study the Japanese production systems. Most of the engineers were relatives of GM executives. They were chosen for the trip as a family perk but were not, in my opinion, the best engineers that GM had. It was November of 1984 and we met the CEO of Toyota and visited two assembly plants. One GMAD engineer on the trip, Ed Klavetch, (not his real name) had demonstrated a positive attitude when he worked for me on powerhouse abatement, but after joining GMAD, the opposite was true with regard to his attitude about spray booth technology.

On returning from Japan, the GMAD engineer's report reflected his lack of enthusiasm for anything he saw in technology. His mood throughout the trip was down and depressing. To see a new country, to meet the chairman of Toyota's board, to discover new devices in your field, to have something good to report to your management, and to discover ways to help your company should have been an exhilarating experience. It was for me, but not for him. What makes a good man change?

An example of this man's application of Godfrey's "no controls" edict was demonstrated in his review of Toyota's spray booth abatement system in Japan. Japanese engineers discovered that activated carbon could be made into a fabric and fashioned into a large circular filter to absorb hydrocarbon emissions. As a fabric, it provided a surface area more than a thousand times larger than the same amount of material in a canister of particles. The collector, placed in the paint booth exhaust stream, absorbed the fumes in the top half of a slow-moving eight-foot diameter fabric wheel, while the bottom half was heated by the incineration of fumes. The VOC emissions were significantly reduced.

The way that Ed questioned Toyota's engineers startled me: "How many men does it take to operate the device?" he asked.

The reply: "None." This question surprised the Toyota engineers; the device was controlled with production shift timers, meaning that it stopped and started automatically.

"How do you know that it is working?"

"Sensors relay performance data to the engineer's office."

"How often do you have to replace the activated carbon filter?"

"It has been operating for three years without change."

"But what is its operating efficiency today?"

The Toyota engineer opened an operation record book about eight inches thick and reported the filters at 78% efficient. "The filter will be used until the efficiency falls off dramatically," he replied.

Toyota's files on the fabric collector had charts and graphs, were very detailed and were separated into sections for easy reference. At GM Godfrey had ordered all records to be destroyed. Also destroyed were all records of engineering data that related to the fixed-bed and fluidized-bed adsorption units tested at the Fremont, California, plant under a legal agreement with EPA. A few years later a division wanted to use carbon adsorption to solve a problem and asked GMAD engineers for their EPA Fremont plant test findings. They had to be told that Godfrey had the records destroyed.

Ed Klavetch, a two-year office neighbor to current CEO Mary Barra, was involved in another scheme to avoid the use of abatement devices. It gives further evidence of his new thinking and the control that GMAD had over him and his future. Prior to its conversion to a waterborne-compliant plant in the early '70s, the Van Nuys, California, plant had serious problems with the emission of odors from the plant's systems. In conjunction with changes to meet the strict emission standards of California Rule 66, a decision was made to install new spray booths and automatic robot spray machines.

The new Van Nuys plant manager, Ernest Schaefer, who had been marked as the best up-and-coming executive, put his reputation on the line. In the end that was costly. He contacted the community neighbors by knocking on each door and saying, "Be patient…everything will be solved with this new system." True, the new system did have fewer discharge stacks on the roof and the neighbors adjacent to the plants could see this improvement. However, the robots in a very small section of the spray booth applied more of the coating than ever before. This concentrated the discharge fumes to a few stacks. The neighbors went up in arms over the increase in odors particularly after promises of improvement. So what was the GM response? The Ford plant in Flat Rock, Michigan, and the Chrysler Jefferson Avenue plant in Detroit had both achieved full abatement of odors. Did GM install odor abatement controls? No, they raised the stacks to make them taller, under the theory that the plume of emissions would pass over the residential areas. But even then, the engineering study was poorly done.

Because of the plume studies that I had participated in with a New York City University professor, I became interested in the Van Nuys problem and reviewed the GMAD wind tunnel experiment. It was obvious that they did not consider how the adjoining high buildings would affect downdraft on the side opposite the residential area. GM proceeded to raise the stacks on the roof again and again, making it possible to see the stacks as you drove over the mountain pass from the Los Angeles basin to the valley. The plant looked like a giant church organ.

It seemed like the stacks could not be raised high enough to eliminate the downdraft, so a meeting was held at the GMAD office with David Campbell, who was vice president of Operations and a Godfrey "yes" man. I sat by myself on a bleacher overlooking the conference table and listened to Campbell say, "I guess we have no other choice but to raise the stacks even higher."

I spoke up: "There are other options."

Campbell shot back, "Yes, Nick wants to consider controls and abatement." He directed his remarks to Leonard "Len" Hostetter, Facility Planning director, sitting at the opposite end of the table. "The carbon adsorption units tested under an EPA agreement at Fremont don't work, right?"

Len replied, "No, the fluidized-bed and fixed-bed carbon units work fine."

"But they cost too much to operate?"

Len replied a second time, "No, the operating cost is about 10% of the installation cost."

"But they won't last long and have to be replaced often?"

"No, they ran for a year or two and only dropped a few percent in efficiency," Len explained. I could have kissed him right then!

But GMAD raised the stacks again. Looking back, it is obvious that it would have been much cheaper to abate than to raise the stacks. However, that was Godfrey thinking at work.

In conclusion, of these three options—enamel, lower solvent lacquer, and abatement—the best approach in my professional opinion would have been to develop lower-emitting lacquer paint and, if need be, apply some abatement. This would have allowed GM to maintain a paint quality advantage over Ford and Chrysler. It would not have required any revision to the paint shop, thereby minimizing capital outlay and costly plant shutdowns.

Viable options were available. The GM Paint Plan was not necessary. Without it, GM would have been free to schedule upgrades to its facilities as product planning, sales, and other economic factors played out. The decision to construct a new plant would come from

*When you came over the San Gabriel Mountains, the stacks on GM's Van Nuys, California, plant had been raised so high that they looked like the pipes of a gigantic organ.*

Credit: GM Public Document

sales projections or the requirements of a new assembly method like the FWD (Front Wheel Drive) platform design. When constructing a new plant, engineers would have time to analyze the new technologies and reach for zero emissions by applying the most promising new technologies, but measure performance and make adjustments *before* the next project. Then, and only then, would the more costly conversion to NSPS, as promised for all plants by the GM Paint Plan, have been required. But engineering logic was not in the driver's seat at General Motors.

# 11

# Joseph Godfrey's Influence

In the business news articles that analyze GM in the 1980s his name is unlikely to appear, but within the corporation, Joseph Godfrey's influence was widespread. When complaints are leveled at the management decisions that came out of GMAD it must be noted that one man was in charge. Why the newly appointed Executive Vice President Pete Estes selected a man in the early 1970s with essentially no vehicle design or assembly plant experience for such a strategic position is a mystery to me. His prior record of leadership at the Saginaw Steering Gear plant was notable, not for great success, but for his failure to correct the flawed O-ring seals, even as warranty costs continued to pile up and customer confidence sank. Tens of thousands of units went into cars and caused cold weather failure until the engine warmed up.

## A Bias Against Air-Conditioning

I first encountered Joseph Godfrey's management style personally in the 1960s. Godfrey was the son of a powerful former vice president of GM. I often felt that this case of nepotism contributed to much of the unquestioned foolishness of Godfrey's leadership. Was he granted immunity from critique because of this connection? Was he driven by a desire to achieve greater things for GM than his father did?

In the late 1960s, Godfrey appropriated funds for the Pontiac Truck Assembly plant—the same plant where GM tried to deny that there were any odor problems—promising the neighbors that they "wouldn't know that the plant was there." It was there that I had my first run in with Godfrey, and it was over the topic of plant air-conditioning. His appropriation requested $6 per square foot of plant for the ventilation systems in the production and assembly areas. The designated office areas were of course air-conditioned. In commenting on the project, I wrote that for $2 per square foot, which amounted to $4 million less for the entire million square foot plant, the production and assembly areas could be air-conditioned and provide the same comfortable conditions experienced by the salaried office workers. To some, that was a shocking statement. Even today, some people believe that on a hot day, blowing outside air at a temperature of 90°-plus on a worker provides a satisfactory work environment and is cheaper than air-conditioning, but I knew for a fact that ventilation was neither satisfactory nor cheap. And I had personal reasons to care.

In 1942, at the age of 13, as I described in Chapter Two "Thinking Differently," on page 11, I worked on the production line at the Ford Motor Company, River Rouge plant complex. In the winter, the manufacturing process kept us warm, but in the summer the heat in the plant was almost unbearable.

At 18, when I went to work for Beaver Tool and Engineering Company and then later for a small tool and die shop, I experienced again that manufacturing facilities could be brutally hot in the summer months.

In the past, GM would hang signs around the building stating to the workers: "We're all on the same team." Meanwhile, the offices of the salaried workers were at a comfortable 72° while the assembly line workers had to deal with temperatures exceeding 90°, sometimes 100°. The air-conditioning at the new truck plant became a personal high priority, but could I convince Godfrey of its importance?

A few years before the Pontiac appropriation, General Motors Institute Professor Leonard Wocholski and a group of Chevrolet engineers, including myself, conducted in-depth studies of four warehouses: two air-conditioned and two that were simply ventilated. We began this task with the support of Dr. Hardie of the design firm of Albert Kahn, and with the encouragement of Jack Cragen, Chevrolet project manager. The group found that the air-conditioned warehouses were much lower in operating cost for any month of the year. The same results were confirmed in an analysis of the energy costs at GM's two WWII bomber plants that were air-conditioned.

Both the plant in Texas and the one in Kansas City were less costly to operate in any month or year when compared to other plants that were just ventilated. More remarkably, the two air-conditioned plants were poorly insulated with large glass outer walls and high ceilings. The Corporate Plant Engineering energy usage reports clearly indicated that even these plants were much less expensive in energy and maintenance costs than those without air-conditioning. There was an obvious reduction in energy consumption when the plant was air-conditioned.

Over the years, at assembly plants, GM had tried to satisfy worker complaints and avoid union walkouts precipitated by excessive summer heat by installing hundreds of individual fans at workstations. These smaller fans were supplemented with roof-mounted air-handling units providing massive amounts of air to the workspace. But blowing tons of 90° outside air over a worker does not compare to small volumes of 55° to 60° air delivered to the lower eight-foot level of the plant. In addition, the studies revealed that the air handling equipment required for the ventilation system actually used more energy than the production machinery and foundry melting pots.

With these studies to support my assertions, I met with Godfrey, who (I believe purposely) mispronounced my name by calling me "Cock-man" instead of "Ketch-man." Godfrey said that he did not believe my statements, but that he would investigate. His head controller accountant worked late that night and reported his findings the next Monday. The accountant said that my figures were not totally accurate, but that it was true that it cost less to operate the air-conditioned plants. So, the Pontiac, Michigan, plant was to be air-conditioned along with every future GM plant. I thought *I won* and took great pride in improving in-plant conditions for all future laborers.

Or so I thought. A few years later I had the opportunity to show off the same Pontiac plant, "General Motors' most modern plant," to Mr. Tanner, head of the Michigan Department of Natural Resources. It was a hot summer day, and to my disappointment, the plant was very uncomfortable. The next day I inquired as to why. The reply was, "Godfrey ordered the thermostats to be at 80° or above. He didn't want the union to think all plants could be air-conditioned." Believe it or not, the union never asked GM to refit any of its existing facilities—they knew the cost of retrofitting would be unreasonably expensive. However, Godfrey's true colors had been exposed.

As a side note, upon reading this anecdote, my daughter, Cynthia, whose husband Anthony has worked as a maintenance supervisor as well as mechanic in several different plants, responded: "This reminds me of something Tony would fight for. There is a difference between a man who has worked as both the laborer and the boss and a man who has always been the boss. The man who has both types of experiences makes a better boss because he is willing to make choices for the good of the people. Men who start at the top without getting their hands dirty do not know what it is to work in bad conditions and feel that their voices are not heard."

## Cost vs. Quality

Joseph Godfrey's ability to implement illogical ideas without critique is difficult to understand. It is theorized that because of his father, he had voting power and had to be listened to. Godfrey's father and his group of former executives had a great deal of influence and some said Estes had to satisfy Godfrey to become the next president of GM. In his position of power, Godfrey issued a number of edicts that, amazingly, still control plant environmental engineers today.

Once Godfrey was in command of GMAD, he imposed his operating philosophy on the entire assembly division. History would show that many of his management ideas were shortsighted. While Godfrey had grandiose plans for himself and GM, there was little consideration of long-term outcomes and a glaring lack of honest cost/benefit analysis in the majority of his plans, whether they were for a small single auto part or an entire multimillion-dollar assembly plant.

Consider the effect on auto parts. Some good resulted, such as uniform assembly standards, with the establishment of GMAD and one vice president over all the assembly plants; but Godfrey's goal was to reduce labor and material costs with little regard for the impact on quality. This disregard for quality angered the division heads like DeLorean at Chevrolet who had been working hard to improve the reputation of their products. Under Godfrey's mandates, Chevrolet was forced to replace a component that had proven reliable with a standardized GMAD part that had failure issues. The warranty costs outstripped any potential savings from standardization.

DeLorean recalls that Godfrey's drive to reduce cost was evident "from the day GMAD took over the Lordstown plant, when it fired 700 workers, many employed in quality control areas. The division also took out the computer quality-defect feedback system."[22] DeLorean goes on to state that:

> The creation of the GM Assembly Division (GMAD) to build almost all GM cars in North America was explained as a move to build cars more efficiently, at less cost. But the 'down time' experience, when the division 'got tough' with the union locals in each plant it took over, cost the company more money than it would make up in savings for as long as a decade.
>
> The company's heavy-handed cost squeeze often forced the parts divisions to push out inferior parts. At one time, the car and truck divisions refused to accept some of these parts because we at least had responsibility for the cars built in our home plants (GMAD built the rest), and we were trying to build them with the best quality possible. The assembly division, however, had no receiving inspection, and so it accepted some of these parts and put them in cars, which, unfortunately, went out with our division names on them. On numerous occasions when touring our parts plants, I saw huge shipments ready for delivery and marked 'do not ship to home plants.' We wouldn't have accepted them. GMAD did.[23]

All of this cost-cutting implemented throughout the division came right at a time when the Japanese were entering the U.S. auto market advertising quality and the reliability of their products.

Another bad idea credited to Godfrey was the redesign of the engine compartment "for ease to assemble." This may have been a cost savings at the production stage, but for the owner and the mechanics, it changed simple maintenance and repair jobs into tasks that were both labor intensive and expensive.

On many of these redesigned GM passenger cars, the engine had to be unbolted from its mounts and lifted just to replace the alternator or transmission cooling lines and sometimes even for access to the spark plugs. Need to replace a water pump? On the new models the air-conditioning condenser, radiator, hoses, fan housing, and numerous other parts had to be removed first. Forget the days of owners being able to repair their own vehicle. GM lost sight of the fact that serviceability was an item of value to many customers. During this same time period in the mid-1970s, Chrysler instituted a policy of "ease of maintenance" and garnered new customers.

And these issues were only the tip of the iceberg. Consider the implications at the plant level.

## "Nothing New!"

In 1973, GM planned to build its first new assembly plant in years in Oklahoma City under Godfrey's direction. Godfrey expressed concern that this plant would be in a new state for GM and that a new product was to be built using all new employees. *There is too much new!* He protested. *Use only well-proven production systems.* His concern was that the workers were not up to handling the latest technology.

Godfrey issued the following prohibitions: there was to be no power-free conveyors (manual transfer of materials between lines), no automatic welders (hand-operated only), and no automatic robotic arms or computerized control systems in the spray booths (hand-held spray guns only). Godfrey also ordered the developers to avoid any technology that was not approved by him. After 10 years of operation, Chairman of the Board Roger Smith ordered the executive committee members to attend the dedication ceremonies and made a huge deal of the events.

Vice President Alex Mair said he would not go and Smith demanded to know why. Mair replied, "If I want to see an old plant like Oklahoma City, I can go across town to the Detroit Cadillac facility that is 75 years old and see the same, if not better, production systems." Mair's comment was sad, but true.

Roger Smith was surprised. He did not know how badly the plants were operating and neither did other top executives. GMAD deliberately kept information from them.

How bad was the Oklahoma City plant and could it be saved? The answer to these questions is clear enough: in 2006, when GM announced its turn-around plan for Wall Street, the Oklahoma plant, a relatively new building, was to be closed. It was an inefficient plant and everyone knew it. How could it have been built so poorly? Godfrey was driven by forces other than a drive for quality and the corporate culture let it happen. Why would he avoid new technology in Oklahoma out of concern for the untrained workers and then propose a plan for paint systems that put those same workers in charge of untested technology?

It must be remembered however that while he was the main policy maker, Godfrey did not operate in a vacuum. In 1978 when the Paint Plan was implemented, Godfrey held the position of "Vice President and Group Executive" of the Body and Assembly Group, but over him were four executive vice presidents, including Howard Kehrl, James McDonald, and Roger Smith and over them was the COB/CEO, the president/COO and the vice chairman. Any one of them could have reviewed and controlled the priorities of the Assembly Division but as already noted a prideful disregard for the interests of the consumer was (and continues to be) characteristic of the corporation as a whole and self-interest drove more decisions than interest in the future of the corporation. All are to blame for GMAD's failures.

## Rule of Thumb Engineering

As evidenced by the Oklahoma plant, Godfrey and his people did not believe in engineering tables and their formulas. The designs for their buildings and systems were largely done by "Rules of Thumb." They took an approach that was customary and practical but it was neither mathematically precise nor based on current engineering evidence. This is a tolerable approach for a backyard tree house, but for a multimillion-dollar manufacturing plant?

Completed after his departure but designed according to Godfrey's priorities, the Detroit Poletown Cadillac plant, the Lake Orion plant, and the Wentzville, Missouri, plant were inefficient and overbuilt. The result in each was 3.6 million square feet under one roof. To this day, each plant has two or three floors underground—like streets or huge tunnels—that are used as air ducts for diluting the polluted paint spray station exhaust with all the other air and for directing the wastewater to a treatment tank far below the ground. The result of this design is over 6 million square feet of constructed space compared to an industry average of 1 to 1.5 million square feet. Prior to construction, in true GM style, there is no record of any debate or questioning of the soundness of the underground ventilation system. No one outside of the GMAD operating group was consulted or given the opportunity to evaluate this unusual airflow plan, nor was there a request for operating data from any similar system in use.

The powerhouse, built at a cost of $75 million, is three times the cost and capacity of what actually is required. This overbuilt powerhouse contains two utility-sized coal-fired boilers costing approximately $35 of the $75 million, whereas one large gas-fired boiler purchased for $5 million would have produced enough steam to serve all of the plant's needs. That's a cost savings of $30 million with the added benefits of freeing the powerhouse management from the issues of coal handling, particulate pollution, and greenhouse gases. Natural gas is delivered via a pipeline, contains no particulates and produces less greenhouse gases when it is burned as compared to coal.

*One of the oversized powerhouses built by GM at great expense in the '80s. It now sits idle.*

Credit: Robert Kachman

How was this powerhouse decision connected to any corporate goals? As further evidence of this folly, today's GM, under the scrutiny brought by bankruptcy, outsources all powerhouse operations and only a few coal-fired boilers are still in use. The steam demand "rule of thumb" was 0.1 pound of steam per square foot of plant and was not based on any actual study of heat gains and losses and steam demand by manufacturing and heating systems. Compressed air supplies for the pneumatic systems and the electrical distribution network were also designed by "rule of thumb."

There is little comparison between those huge overbuilt facilities and new assembly plants like the one in Lansing, Michigan, which is built in the shape of a "Y," thus permitting the plant manager to see every operation from the facility's center. In Godfrey's plant, on the other hand, a maintenance worker sent to a repair job might not be seen for days. Not only are the underground levels cavernous, but also the main floor is inefficiently oversized.

If the Chevrolet Central Office construction engineers, who had previous assembly plant construction experience, had been given an opportunity to conduct an engineering evaluation of GMAD's design, most construction faults and excessive services could have been avoided. But the newly formed GMAD was too powerful, and nobody dared to question the bringing of several train cars into the plant to avoid weather, a powerhouse three times the required size, or the ridiculous underground air handling and wastewater systems. Most felt they were being good company men by not questioning anything—engineers that did were pushed aside and isolated.

Alex Mair was quoted as saying, "You could find space for three assembly lines in the three new plants, not just one."

## LIMITED RESEARCH—NO ABATEMENT

Too many of Joseph Godfrey's leadership decisions were costly to GM but the pinnacle of Godfrey's folly was in the Paint Plan. The most significant contributing factors were his clamp on paint research and his edict that abatement could never be an option.

In addition to cost cutting, one of the first things Godfrey did upon his 1974 appointment to lead both GMAD and Fisher Body was to convince Pete Estes to transfer the Tech Center Manufacturing Staff

Paint Research and Development Group to the GMAD Operating Staff, effectively placing it under Godfrey's control. The transfer of paint research was a surprise to many because never before nor since, to my knowledge, has any tech center development or research group been placed under the complete control of operating plant management. Research and daily operating people have divergent objectives. Research, seeking to discover the best solution, must actively seek out and consider all options without interference from plant people, who are predisposed to suggest solutions that vary little from current protocol. This is what occurred.

Godfrey immediately curtailed the development of a promising alternative in powder-paint as well as halting research on high-solids enamel and the spray equipment needed to apply them. His bias toward the exclusive use of lacquer was not hidden. Additionally, gag orders regarding research went into place. Speeches, reports, and press releases about paint development were henceforth suppressed and blocked from review not only by outsiders, but even by top management.

The continued use of lacquer was all but guaranteed on passenger cars throughout the late 1970s. When GM was under mounting pressure from the regulatory agencies and was forced to make a move, Godfrey endorsed the Paint Plan with its proposal of a yet undeveloped paint technology. It is theorized that Godfrey didn't want any questions from management about the Paint Plan. Had he not limited the free flow of information, some smart executive following GM's paint research might have discovered that there were better alternatives and that the GM commitment to the EPA was unnecessary and foolhardy. The whole Plan might have been avoided.

But as discussed in detail later, the Paint Plan was successfully sold to upper management as the only alternative for confronting EPA regulation. It was accepted, with skepticism, by David Hawkins in the EPA but then little was done within GM to implement it within the time constraints that it established. Rather than pushing the process along, Godfrey's man Dave Campbell was dragging his heels on completing the conversions. In plant after plant, little was done to meet the compliance deadline and no evidence exists of an intentional effort to provide accountability. Instead there was continued pushback against the regulators.

Why was there no follow through? As mentioned earlier, it is theorized that Godfrey was awaiting a change in the U.S. presidency hoping that the EPA standards would be relaxed under Reagan, who had run on an "industry friendly" platform. In fact, Reagan's secretary of interior and EPA administrators did do industry's bidding even beyond the law thus ending their careers with heavy fines and even jail time.

Godfrey gave evidence that he was convinced that GM should continue the use of the existing lacquer paint as long as possible because it provided a repair cost advantage over the competition who was applying enamel paint. Yet he didn't push for development of lacquers with lower emission rates. He also wasn't too concerned with rust problems as a downside of lacquer because in his mind, rust is good for business since it encourages new vehicle purchases. He was actually heard to say, "Obsolescence is a good thing for GM."

Godfrey's clamp on paint research and his aversion to abatement can be documented in the early '70s, but it is after an industry cocktail party in 1977 that his policies were cemented in place. The GM Fremont, California, plant set the stage.

The topography and air patterns of California, combined with its dense population, have historically put the state in the lead for pollution regulation. To meet California Rule 66, which forced reduction of pollutant air emissions, GM converted its auto paint from lacquer to a waterborne coating and in doing so established the NSPS (New Source Performance Standard) and LAER (Lowest Available Emission Rate) standards for the automotive industry. Whenever a company demonstrated the feasibility of a lower emissions standard, that standard became the EPA's benchmark for the entire industry. Not long after the summer of 1977, when the EPA issued the new standards based on the GM plant, Godfrey attended a cocktail party of executives from the domestic auto companies. They raised hell over the establishment of the stricter EPA standard by the Los Angeles plant. Godfrey's pride was injured. Beating Ford and Chrysler was generally at the top of the GM incentive list, but on the issue of resisting government regulation, the industry was united. Seemingly as a result, the following edicts were issued by Godfrey and they became as law within the corporation for years to come:

- Abatement of paint booth emissions with stack add-on controls was forbidden.
- In implementing the Paint Plan, no new environmental standard was to be established:
    - Developments in paints and spray application systems must not reduce emissions beyond current EPA standards.
    - State and federal agencies may not be provided with technology leading to a new pollution standard.

To ensure Godfrey's way of doing things were followed, he appointed a man devoid of any creativity to head the paint development department. And as detailed in the previous chapter, after the Fremont plant was closed, Godfrey ordered all records of performance from the abatement system installed in that plant to be destroyed. A phrase often heard was, "I don't want to see a chemical plant (meaning abatement) on my roof." Godfrey's new paint man met Ford and Chrysler paint managers in the early 1980s at the Oklahoma City plant. GMAD selected this plant for our competitors to see because of all the old technologies in use. These managers from Ford and Chrysler were very happy to know that GM was so far behind.

## The Epitome of Inefficiency

If they were unimpressed by the plant in Oklahoma City, I wonder what they would have thought of the inefficiency of the five plants that were built between 1981 and 1987. In addition to being oversized, the Hamtramck, Lake Orion, and Wentzville assembly plants, along with the Shreveport, Louisiana, plant (1981) and the Fairfax, Kansas City, Kansas, plant (1987), were designed with cavernous underground tunnels to deal with the polluted water and air that comes out of the spray booths.

The spray booth water treatment and sludge handling system is three to four stories below grade and has been described as category four whitewater rapids. Imagine the construction and maintenance costs of such a design and the inflexible barrier it erected to future changes in the plant design.

In every other plant and in every other company, paint exhaust is fed out of roof exhaust stacks and even water treatment systems are above ground. These particular GM plants were built with an underground city below the main floor. Streets acting as ductwork were designed, some to carry away the paint sludge and others to mix clean air with the concentrated polluted exhaust from the paint spray booth. In order to pull all the air from the tunnels under the spray booths through this underground system, the design requires a large underground room with a dozen or more two hundred horsepower fans. Together these fans combine the air from several "streets" into a cavernous hall—much like the Pantheon in Rome—and then send it out one giant exhaust stack. To maintain the needed level of exhaust performance in the spray booths, more fans and increased electrical power had to be added after the initial construction. The underground room is large enough to drive a herd of elephants through it and the enormity of the roof exhaust stack is obvious when viewed in Google Earth. Amazingly, this elaborate system was built to dilute, not reduce, the pollutants as a means to meet regulatory standards.

In early 1980, American Motor Company engineers learned that GM's new assembly plants had a "revolutionary" way of ventilating the spray booth exhaust in an underground duct system with only one discharge stack. They could not determine why GM's system was better than the traditional system. To find out, their plant engineer Richard Schenkel called me, asking, "Why?"

*The standard ventilation system, which utilizes a separate set of stacks for each paint booth.*

Credit: GM Internal Study, 1978

This is a question I had posed to myself. From an engineering perspective, the design was ludicrous. The only answer that was feasible was that the underground system had been proposed as a way to comply with Godfrey's edict to avoid abatement controls. Controls would have not only cost a fraction of the underground air system, but also effectively removed pollutants, not just diluted them. Dilution was not a solution to pollution!

"Because we're nuts!" was my response. "We're combining the polluted paint air with as much clean air as is available just to avoid pollution controls. It costs tens of millions more and controlling the spray booth air flows will be difficult. It may not work. Dick, don't do it."

He thanked me and hung up.

A first-year engineering student evaluating this enormous air system would have known that these plants should not have been built, but under Godfrey's influence, not one, but eventually five plants were built with this crazy system and put into operation. It was typical of operations at GM that the engineers at American Motors were asking questions never discussed within GM.

Before retiring, Leonard Hostetter, the director of Operation Engineering Services for the CPC Group responsible for the construction of the new plants, told me he was sorry for not listening—the plants should have been out of the pollution business for the good of the company. It was nice to hear, but it came much too late.

So in Joseph Godfrey we have a man who resisted the installation of air-conditioning even when it was proven to be more cost-effective, who resisted improvements in technology, who resisted reduction of paint emissions, and who resisted the application of good engineering practices. Then in 1977 along comes a plan, unconnected to customer demand, unconnected to the corporate production plan, unconnected to return on investment figures, involving a voluntary commitment to the EPA for reduction in paint emissions from all 39 GM plants achieved by the total overhaul of every plant's paint system at a conservatively estimated capital cost of billions. Why would a man who was normally so resistant suddenly push GM into a commitment to a modernization that was so extensive, expensive, and unrealistic in its promised 10-year time frame?

Credit: Robert Kachman

*A monument to poor engineering, this abandoned stack connects to a cavernous underground room with concrete tunnels that used giant fans to draw air from all the paint booths. The entire system has long been idled but there is no feasible way to remove it.*

# 12

# The Development of the Paint Plan

The Godfrey Paint Plan evolved from a document known within GM and the EPA as "the yellow book." The yellow book contained GM's formal response to the EPA's stricter requirements under the amendments to the Clean Air Act. Interestingly, all the copies of this original plan were eventually destroyed and to my knowledge none is in existence today, but internal memoranda and letters from EPA officials speak of the yellow book and its plan for the paint shops in the GM assembly plants around the country. The yellow book presented a timeline for the conversion of all its assembly plants to a new (yet to be developed) paint system that would meet not just the RACT standard but the more restrictive NSPS standard. The New Source Performance Standards held forth the most restrictive industry emission rates and would generally require a more costly system. GM made a legally binding commitment to the EPA that, utilizing new technology, they would go beyond the requirements of the law for existing facilities. They committed to using a new paint that had a primer coating that was still under critical development and spray apparatuses and robots that were untested.

The original conversion plan called for 39 major paint shops, 33 of which were in 31 assembly plants with 6 in Fisher Body plants, to be modernized between 1978 and 1987. The paint for the color topcoat, still in development, would not be selected until three years into the program, leaving only seven years for implementation to meet the compliance deadline. Ignoring the uncertainty of GM's ability to fulfill the commitment, it was a great offer to any environmentalist and to the environmental agencies. GM saw it as a way to satisfy both the state and federal agencies in one comprehensive plan, thus avoiding piecemeal negotiations.

The chart entitled "GM Conversion Plan 1978–1987" shows that the first step at each plant would be to ensure that electrodeposition, ELPO, was in use for the application of the primer paint. The next step would be to use 27% lacquer and the final step was to put the new technology in place for the application of GM's new enamel paint that would have an emission rate equal to the low emissions of the waterborne enamels. The undeveloped new paint was hence known as "WBeq" for "waterborne-equivalency." Using the logic applied by the steel industry, GM did request from the EPA an extension to 10 years for compliance. An inch-thick detailed internal report dated November 1, 1978, made the case that there were not enough capital funds or contractors for 31 plants to shut and rebuild their paint shops in five years. It proffered a plan for conversion to new paint systems over a 10-year period. Given the rationale and scope of the offer, the EPA accepted the time extension, but within the 10-year time frame individual compliance deadlines were established for each of the 31 plants.

Contrary to common business sense, the compliance date for each plant was set with complete disregard for the product plan for that plant and was determined solely on the basis of the air quality of the area surrounding the plant. Ludicrous from the perspective of corporate profits, those in more polluted areas were scheduled to be upgraded first and those in less polluted areas would be converted later.

As for the total cost of the plan, the estimates given in 1977 were that the conversion of all the plants would cost GM in the neighborhood of $4 billion. Although it was deemed necessary to meet environmental regulations, that was *$4 billion* earmarked for an upgrade that most customers would never notice. In short order it became apparent that the $4 billion was an optimistic estimate. The actual cost when calculated 10 years later was 10 times that amount.

No other automotive company in the world, past or present, in the best of financial health, would ever commit to a radical modernization of all its product assembly plants in 10 years and make it a self-induced legal requirement without regard for changing consumer demands and economic cycles.

Yet Group Vice President Joseph Godfrey's plans were even bolder and costlier. His idea was to use the funding of the Paint Plan as the justification to close many of the older factories, build new assembly

**GM CONVERSION PLAN 1978 - 1987**

**DECISION DATE** (arrow pointing to 1980)

| STATE | PLANT | 1978 | 1979 | 1980 | 1981 | 1982 | 1983 | 1984 | 1985 | 1986 | 1987 |
|---|---|---|---|---|---|---|---|---|---|---|---|
| CALIFORNIA | FREMONT CAR | 17%DL | | 27%DL | COMMITMENTS BEYOND 1980 TO BE RESOLVED | | | | | | |
| | FREMONT TRUCK (SHUTDOWN) | | | 40-45%SE | | | | | | | |
| DELAWARE | WILMINGTON | | | ELPO/WPS | 27%DL | | | | WBEq | | |
| GEORGIA | DORAVILLE | | 17%DL | | ELPO/WPS | 27%DL | | | | | WBEq |
| | LAKEWOOD CAR | | 17%DL | | 27%DL | ELPO/WPS | | | | | WBEq |
| | LAKEWOOD TRUCK (SHUTDOWN) | | | ELPO | | 50%SE | | | | WBEq | |
| KANSAS | FAIRFAX | | 17%DL | | 27%DL | ELPO/WPS | | | | | WBEq |
| MARYLAND | BALTIMORE CAR (SHUTDOWN) | | | 17%DL | 27%DL | ELPO/WPS | | WBEq | | | |
| | BALTIMORE TRUCK | | | | ELPO | 50%SE | WBEq | | | | |
| MASSACHUSETTS | FRAMINGHAM | | 17%DL | WPS | | 27%DL | | | WBEq | | |
| MICHIGAN | BUICK MOTOR "A" & "B" SHUTDOWN B.C. | | 17%DL | | ELPO | 27%DL | | | WBEq | | |
| | FLINT NORTH & SOUTH | | 17%DL | | ELPO/WPS | 27%DL | | | WBEq | | |
| | CADILLAC MOTOR | | 17%DL | | 27%DL | ELPO/WPS | WBEq | | | | |
| | FLEETWOOD | | 17%DL | WPS | 27%DL | | WBEq | | | | |
| | (SHUTDOWN) OLDSMOBILE MOTOR | 17%DL | | | 27%DL | ELPO | | | | WBEq | |
| | LANSING "A" & "B" | ELPO/WPS | | | 27%DL | | | | | WBEq | |
| | | 17%DL | | | | | | | | | |
| | PONTIAC MOTOR | | | | 27%DL | | | WBEq | | | |
| | PONTIAC "A" & "B" SHUTDOWN (ORION) | | | | 27%DL | ELPO/WPS | | WBEq | | | |
| | FLINT TRUCK #1 | | | | ELPO | ELPO(S/M) | WBEq | | | | |
| | | | | | | 50%SE | | | | | |
| | FLINT TRUCK #2 | | | | 50%SE | ELPO(S/M) | WBEq | | | | |
| | GMC #3 (VAN) | | | | 50%SE | | | | WBEq | | |
| | GMC #6 (TRUCK) * SHUTDOWN (GM T 400) | | | | 50%SE | | | | WBEq | | |
| | WILLOW RUN | ELPO | ELPO(S/M) | WPS | | WBEq | | | | | |
| | | | 17%DL | | | | | | | | |
| MISSOURI | LEEDS | | 17%DL | WPS | 27%DL | | | | | | WBEq |
| | ST. LOUIS CAR (SHUTDOWN) WENTZ | | 17%DL | | 27%DL | ELPO/WPS | | WBEq | | | |
| | ST. LOUIS TRUCK | | | | 50%SE | ELPO | | | WBEq | | |
| | ST. LOUIS CORVETTE (SHUTDOWN) | | | | | | | WBEq | | | |
| NEW JERSEY | LINDEN | (GM 25) | WPS | 17%DL | | 27%DL | WBEq | | | | |
| NEW YORK | TARRYTOWN | | | | 27%DL | ELPO | WBEq | | | | |
| OHIO | LORDSTOWN CAR | | WPS | 27%DL | | | | | WBEq | | |
| | LORDSTOWN VAN | | | | | 50%SE | | WBEq | | | |
| | NORWOOD | | 17%DL | | WPS/WBE | | | | | | |
| TEXAS | ARLINGTON | | 17%DL | WPS | ELPO(S/M) | 27%DL | | | | WBEq | |
| WISCONSIN | JANESVILLE CAR | ELPO(S/M) | 17%DL | WPS | 27%DL | | | | | WBEq | |
| | JANESVILLE TRUCK | | | ELPO(S/M) | | 50%SE | | | | | WBEq |

WPS - WATERBORNE PRIMER-SURFACER
DL - DISPERSION LACQUER
(S/M) - FRONT-END SHEET METAL

SE - SOLVENT ENAMEL
WBE - WATERBORNE ENAMEL
WBEq - WATERBORNE ENAMEL OR EQUIVALENT

Credit: GM Internal Study, 1978

*The actual 10-year Paint Plan as it was presented both internally and to the EPA. Note that a decision regarding a new paint application—one that would meet compliance requirements and provide a world class finish—was promised to be made within the first three years, leaving only seven years for thirty plus plants to be modernized with the newly developed paint system.*

plants on green field sites, and at the remaining plants, modernize the entire paint shop with small sheet metal press plants attached. All of these expanded ventures had associated costs, which were not considered in the original Paint Plan.

Looking back, an undertaking of such magnitude should have required a great deal of debating, exploring all options, and the consideration of all obvious and hidden costs by addressing the possibilities of future economic conditions, inflation, the cost of production downtime, future product introductions, plant closing costs, and their site cleanup, etc. before the Plan was approved. Even an automotive modernization program of much smaller magnitude, such as a model change or one involving the temporary closing of one or two plants, would normally take many months or years to evaluate, but the Paint Plan received minor scrutiny. Why would engineering and financial analysis not be applied to the biggest project the company ever planned?

According to Saturn's third president, R.G. Le Fauve, the promises of the Godfrey plan were as follows:

1. GM would be the first to implement paint advances that were coming to the whole industry.
2. GM vehicles would have a world class finish, superior to the competition.
3. The problem of staying in compliance would be eliminated by being the industry pacesetter.
4. The many community odor complaints, along with the legal suits, would be resolved.
5. GM would have no trouble obtaining permits.
6. The Paint Plan modernizations would be so technologically advanced that this would be a "once and done" expenditure that would free up future funds for product development.
7. By setting a new standard at all their plants, *what had been NSPS would become the industry RACT* and Ford and Chrysler would be forced to make similar capital investments and it would take years for them to catch up to the advances at GM.

Most of these promises were based on unsupported assumptions, yet they sounded so promising that people accepted them without scrutiny. To my knowledge, only one executive seriously questioned the Plan. Roger Smith, at that time the head of General Motors' financial staff, thought a relatively easy conversion from lacquer coatings to the enamel paint used on all of General Motors' trucks and used by all other industry members would satisfy the regulatory demand and be a less expensive program. He was right.

## Roger Smith Signs Off

I recall that day in 1976 as if it were today. Someone from Roger Smith's office called asking me if Smith could meet with my subordinate Dave Aldorfer on Wednesday morning. Dave believed in the Paint Plan and GMAD management wanted him to run the permitting program for the Environmental Activities Staff without my interference. Aldorfer coordinated the yellow book and, in many ways, began working more for GMAD than for the EAS.

At 8:00 a.m., Smith was sitting in the EAS conference room saying he had met with Joseph Godfrey across the track (the other side of the GM Technical Center) at the Fisher Body Building at 7:00 a.m. They had an interesting exchange of ideas.

Smith spoke of options other than building and modernizing all the lacquer paint shops as well as the enamel truck plants. He questioned the need for an extraordinary expenditure when a conversion to enamel paint would satisfy the RACT compliance standard. Smith was right on track: the truck plants were essentially in compliance, so why create more headaches? Roger Smith was not technical at all; for him to talk about this option, it could be assumed that he had to have had conversations about Godfrey's regulatory approach with other GM staff or perhaps executives outside the company.

I can personally recall an exchange between Godfrey and Roger Smith about enamel paint. Initially Godfrey wanted to stick with lacquer. He and Smith walked down to the parking garage where a beautiful large Ford Lincoln Town Car sat as a demonstration of an enamel finish. Godfrey said, "Look at that! Look at that!" as he pointed to spots on the vehicle's paint.

But Smith later told us that he could not see what Godfrey was trying to point out and just replied, "Ford sure sells a lot of these."

Smith said that Godfrey wanted to keep lacquer because with lacquer GM could quickly spot-repair dents and scratches while our competitors, using enamels, had to redo whole doors or fenders, a process known as panel repair. If forced to discontinue the use of lacquer, Godfrey wanted all new facilities.

Well into the meeting, Smith turned to Vice President Chuck Katko, who was filling in for Godfrey, and asked, "Now tell me if it is really necessary to convert all the plants to waterborne and build new facilities."

Katko, speaking softly, responded, "Yes, it's better than using the existing systems."

Smith took out a large orange fountain pen from the inside pocket of his jacket, unscrewed the top and started to write in his spiral notepad.

I thought, *My God, this guy is great! He's going to hold them accountable and as failures occur, he will surely call them on the carpet.* But Roger Smith never did.

The only other hint of deliberation about the Plan came in 1977 with a visit from Vice President Pais of the Financial Staff who said, without questioning any point of the program, that $4 billion was very doable for GM. Like Godfrey, Pais was the son of a former vice president with little, if any, manufacturing paint system or environmental law experience.

Smith was right to question the necessity of the Plan, but finally went along with Godfrey, probably because of the promise of putting GM far ahead of the competition for some time to come. It was assumed that once the Plan was completed the major appropriations for these plants would stop and money could be spent on the product rather than the plants. I believe both Roger Smith and Joseph Godfrey had visions of themselves as the corporate hero who was leading GM back to worldwide dominance and prominence, and corporate management followed along.

Dave Aldorfer's memo dated April 24, 1978, confirms this line of thinking. Aldorfer writes, "Our competitors are working on higher solid enamels to meet RACT requirements. Higher solids can be used within their existing enamel facilities without major expense. On the other hand, for GM to convert to enamels would require major facility modifications. The Corporate Plan is based on making the expenditures only once and, hopefully with the additional developmental time, to accomplish emission levels not requiring add-on [abatement] controls and future reductions."

## An Outside Perspective

Consider also the observations of John Calcagni of the EPA. In the course of cataloging my observations regarding the GM Paint Plan, I had several conversations with my good friend. At my request Calcagni put into writing his own perspective on the development of the GM yellow book. As an outsider, he was making the same observations about corporate motivations that I was from the inside. It is fascinating to read how transparent we were about our loathing of Ford and Chrysler and how clearly GM decisions were made based on pride, not service.

> I transferred to OAQPS [Office of Air Quality Planning and Standards] on July 1, 1978. One of my first assignments was to negotiate a RACT compliance schedule with GM. I was not at the initial meeting with David Potter, a vice president at GM, but as relayed to me by Jim Berry, GM recognized it faced a difficult problem. Its paint shops were lacquer based and used much more solvent than its competitors whose paints were enamel based. GM believed that lacquer provided a superior finish and gave them a competitive advantage in sales as well as made repairs much easier. However on a lbs./gallon solids applied basis GM was emitting about 125 lbs., while enamel systems were nearer 35–45 lbs., while the GM waterborne plant, upon which the CTG [Control Technology Guideline] limit was based, was about 19 lbs. [*sic*]. Lacquer would never be RACT. [Author's note: a false assumption of many at the time.] If GM conceded that enamel with some improvements in transfer efficiency was RACT then it would have to revamp its painting program while its competitors would essentially get a free ride. It also had to explain away its existing [Van Nuys, California] waterborne plant. While a challenge, if GM could leapfrog the

> industry and develop a world-class coating that met the emission limits, it would put its competitors at a significant disadvantage. That was the theory behind the GM proposal.
>
> GM was adamant that it did not want to convert to waterborne paint even though they knew it provided a quality finish. One of the reasons given was that it would require them to air-condition all their paint shops to maintain color match, *which they said would be prohibitively expensive.* [Emphasis added because I had proven to Godfrey that this statement was untrue.]

In this 2009 statement titled "Development of the Yellow Book – My Perspective," Calcagni accurately described the GM thinking at that time. His observations confirmed our corporate vanity and our corporate ignorance. His statements demonstrate the way in which false assumptions, specifically in this case in regard to advances in lacquer and the expense of air-conditioning, were accepted without examination. The top leadership in any business needs to have sufficient knowledge and experience in the core functions of that business to recognize the difference between facts and assumptions and ask for hard data when assumptions are being presented as the impetus for major business decisions. Consider the outcome of building a bank business around the assumption that housing prices will climb indefinitely.

It is generally the nature of engineers to require hard data and to test assumptions. This had been my training. At the start, my expressed opinion was that the Plan was doomed for failure and was an unnecessary commitment. But people who are enjoying the hope of a sunny day don't want to hear about the storm clouds over the horizon. Because of my opposition, I was not invited to the official signing of the agreement between the EPA and GM in Chicago.

Within the EPA, David Hawkins, who had been the one to attach strict numbers to the industry guidelines, was also opposed to acceptance of the Plan but was eventually convinced by Walt Barber, director of the OAQPS, to go along with it. Calcagni recalls, "Dave was upset when told of it...It was his experience with steel and other industries that they would bargain for time, do nothing and use the time to go to Congress and the courts to extend the schedules further. He did not trust GM to follow through."

## The Fallout

History shows that both Dave and I were correct. GM behaved exactly like the other industries and internally the promises connected to the Paint Plan were not only unfulfilled, but the reverse occurred. Previous odor complaints remained and new ones were created. Our neighbors were not happy with us. The finish on the vehicles was far from world class and, as cited in Keller's book, Roger Smith had to admit that billions had been spent on capital improvements with negligible increase in market share. True to Dave's predictions, as a result of noncompliance with the (self-imposed) deadlines, the EPA was forced to bring suit against GM and to impose fines. The GM legal staff had to hire more attorneys and pay outside counsel to handle the numerous cases. Additional millions were spent on the resulting legal fees alone.

The plan was to outdo the competition. Instead, it placed the company at a great disadvantage compared to its competitors because of the enormous outlay of capital with no resultant increase of sales. The Paint Plan was a financial disaster for GM. It cost GM billions and epitomized all that was wrong in the GM culture leading up to the bankruptcy.

Could the Plan have been avoided? Absolutely! GM never tried. Consideration of other options was never discussed by the Emission Review Board (ERB) or any other body with authority. Godfrey, for reasons discussed, decided the best course for the company should be to modernize the assembly plants as quickly as possible, thereby leapfrogging the competition. He and his staff made a convincing case but it lacked so many important details.

Those who rushed into the Paint Plan must not have read or understood the Clean Air Act (CAA). The regulations imposed on industrial pollution sources were based on sound engineering principles with reasonable cost-effective control standards. As a corporate environmental engineer, I had no disagreement with the CAA. Cities with polluted air needed immediate abatement of emissions to protect public health and the document concerning

"Available Control Technology" (ACT) gave evidence that the government standards could be achieved. GM had at least two pathways to compliance available to them.

## Alternatives

The most sensible step was to have all existing sources install RACT controls that were practical, already in use, and proven to be cost-effective and activate them as soon as possible. GM could have continued with implementation of the new ELPO primer paint—stayed with lacquer topcoat paint but raised the solids level, as two of their paint suppliers were offering to do—improved spray efficiency, and finally added some abatement on the exhaust stacks. This would have allowed them to implement new technology as it was proven and to do so in a way that coincided with planned product changes.

In consideration of the fact that all of the truck plants were already in compliance, all of these steps could have been completed within a five-year time frame and at a reasonable cost, achieving an 80% reduction of emissions. In my mind it was foolish at best to reach for 90% reduction of emissions. I knew that to do so would take more than 10 years. I'm still amazed today that management, the GMAD operation group, and most of my department jumped at the opportunity to promise the higher reduction level, going in this one case above and beyond what the EPA required, while fighting every other environment regulation and vehicle emission and safety proposal by the EPA.

The second alternative pathway to compliance was to take the steps detailed above, implement those that were easily achieved, and then if the plants still exceeded the emission level that was RACT for the industry, ask the EPA to grant a separate RACT number specifically for GM. As discussed earlier, a corporation with such a large economic footprint could request exceptions, and I was sure we could achieve a reduction acceptable to the EPA.

GM's influence was exactly what worried David Hawkins, EPA administrator, who to this day believes granting approval to the Paint Plan was the biggest mistake of his career.

# 13

# Everything That Went Wrong

In many ways, the Godfrey Paint Plan was doomed to failure before it began. Its advocates overlooked or deliberately ignored so many important factors. Issues within the corporation then served to magnify its flaws and thus the GM downward spiral began.

The only factor external to GM that contributed to the failure of the Paint Plan was an economic downturn that occurred in the 1970s. The energy embargo hit GM profits hard. Beginning in 1973, nationwide economic downturns were occurring every few years and many GM products were not selling well, yet the Plan was approved in 1976–1977 based on the supposition that the company was indomitable and would still be in great shape for some time to come. GM pride clouded judgment. What a mistake! This external factor in combination with internal attitudes created a perfect storm.

In the mid-1980s when Roger Smith was leading GM in the capital acquisitions of Electronic Data Systems (1984) and Hughes Aircraft (1985), General Motors' market share was falling, new products were not competitive, the new paint systems were already full of problems and management was burdened with a costly commitment they could not meet. I can attest to the fact that there was palpable stress within the organization during these years. But we could have been prepared. History shows that it is not always easy to predict economic downturns, but it is possible to have a business plan to turn to in the case of a down market. Pride, lack of foresight, lack of oversight, and lack of coordination formed a fountainhead of financial loss.

A memo reporting on the presentation of the Paint Plan to the EPA states, in part, that the Plan would provide a means of maintaining control of facility planning. But control by whom? The Plan may have prevented the state agencies from setting an external compliance timetable but internally it pitted the environmental staff against the product planners. When the two plans didn't match and no effort at coordination was made, the clash presented two possible outcomes. Assembly plants could follow the business plan and delay compliance, or follow the Paint Plan and face temporary closures. The delays in implementation indicate that the product people generally had their say. In time, half a dozen law firms were retained to aid GM Legal Staff in legal battles over plants that received extensions but still didn't comply because compliance might affect the production of the product and sales, i.e. the business plan.

A typical paint shop includes areas for cleaning, applying ELPO, spraying, and baking the paint, as well as treating the wastewater from the process. All of these elements occupy about half the floor space of an assembly plant, representing a sizeable component of the assembly process. Shutting this section down for remodeling stops production. To this day, I ask myself why any auto company would volunteer to shut down one of its assembly plants in the midst of profitable production in order to change the paint applied to the vehicles. Where was the foresight? Some of these plants were actively churning out over 200,000 well-received vehicles a year and generating billions in revenue and profits. A plant that is shut down is a source of expense, not revenue. Scheduling three or four plants per year for paint shop closure with no consideration for possible economic downturns, periodic energy crises, or changes in consumer tastes is shortsighted. But ignoring the Product Business Plan places the offer to the EPA among the most ridiculous business decisions any company has ever made. There was a constant battle of priorities and the product planning was underfunded because of the constant drain on funds imposed by the Paint Plan. In fact, the plants were losing money, not making money.

Good corporate governance and traditional industry practices only allowed a facility to be shut down and modernized or radically changed when the product had to be remodeled, the economic forecast had met certain standards, and/or the facility was outdated and required updating with new, more efficient production equipment. *The changes to the paint shops could have been done in conjunction with other plant upgrades.* The Paint Plan had no correspondence with the GM Product Plan. The Paint Plan actually scheduled the rebuilding of the assembly plants based solely on the ozone air quality in the plant's host city, with the East Coast plants retooling first and the plant in Atlanta, Georgia, last. Ironically, while Atlanta had the least ozone issues in 1975, today, pollution in Atlanta is as bad as in Los Angeles.

When vehicle sales began to fall, new vehicle programs had to be introduced into a product business plan that was completely disorganized. Would any well-managed company commit to closing all of its assembly over 10 years to change the paint systems without coordination with the product plan? Not any company, just GM. And the North American operations suffered.

## Undeveloped Painting Technology

Remember Aldorfer's memo stating that the Paint Plan was "based on making the expenditures only once?" That was not realistic and proved to be entirely misleading. There are two components to painting: the paint and the application system. At the time that GM made its Paint Plan commitment to the EPA, those in management failed to grasp that neither the optimum paint formula nor the paint application systems were fully developed and ready for use. Without consultation with any experts, and with no in-house research being conducted, assumptions were made that GM's paint suppliers and spray system vendors could quickly provide the advances needed to meet the agreed upon deadlines. But the suppliers had no financial incentive to meet the EPA deadlines. And when those deadlines arrived and the pollution levels were still too high, or as in some cases, no change had been made to the paint process at all, it would be GM, not the suppliers, who would have to pay the million dollar fines for noncompliance.

From the use of a brush in the human hand in 1900 into the Paint Plan era, advances came in paint application in the form of increasingly sophisticated spray guns and then robots. The removal of as many workers as possible from inside the spray booths and their replacement with computerized robotics was a major Paint Plan objective. Roger Smith wanted people-less factories and paperless offices. Technology was coming of age.

The traditional handheld spray guns were to be replaced with high-speed rotating bell nozzles—new technology, untested, and extremely expensive. Remember that according to the Paint Plan conversion charts and Godfrey's internal memos, a selection of the topcoat paint and its application method would not be settled until 1980, three years into the Plan. In 1987, at the deadline for the Plan's completion, these paint systems were *still* going through major changes and *improvements that would continue for 20 more years,* through to 2008. In fact, the development of paint apparatus was exponential until the year 2008. Coating materials, primer surfacers, etc., also changed drastically for the first 30 years beyond the start of the Paint Plan. The Paint Plan never had a chance of succeeding because of the time limit it placed on technological advances. New developments were rushed into production before they were thoroughly tested and ready.

The new paint shops contained the most advanced robotics and the latest innovations. They used the best mechanical hardware and electrical components available. But the technology was still young.

Automotive painting is a complex process. The robots had to adjust the proximity of the spray gun to the car body according to the color of the paint, the vehicle shape and size, and the speed of the production line. The mini-bell rotating disc paint spray guns had to be computer adjusted for each distinct color coat because of material and solvent content differences. The line pressure, and on some systems the temperature of the air jets, all required constant adjustment in order to optimize spray efficiency. The computer programming to coordinate all of these functions was in a constant state of flux, frequently stopping the entire production line. With the new robotics, instead of paying blue-collar laborers, GM was paying white-collar programmers.

Throughout the 10-year program and beyond, additional funding for major modifications to the robotic paint system was necessary. Sometimes the requests came more than once at the same plant, and such requests continue to this day. In 1980, while watching the

Orion plant's paint robots paint the spray booth walls while missing the vehicle, Bob Phillips overheard GM President Pete Estes, ask Joe Godfrey, "Will this work?"

Godfrey replied, "I don't know"—two years after the commitment to the EPA.

By 1983, halfway through the Paint Plan schedule, the new paint systems were failing. They were failing to produce a quality finish, failing to reduce the cost of painting, and failing to eliminate environmental permit issues and community odor complaints; in fact, all the new plants were out of compliance. As stated, the paints, robots, spray guns, and the computer programs stayed in development until 2008.

In 2008, GM invested $330 million to upgrade the Saturn paint system, originally built in 1990 for an estimated $750,000,000. This plant was not burdened by Godfrey's underground tunnel system; at the wise discretion of LeFauve, his engineers had designed a spray booth air recycling system that could better accommodate paint formula changes. However, even this was not enough to head off the need for future upgrades.

In selling the 10-year Paint Plan to management, Godfrey promised it would be a one-time cost to modernize each paint shop. Sadly, many of the modernized paint shops like Saturn's are closed today and the paint systems that survived closure are still being modified as vehicle programs change. A GM news release dated January 28, 2013, announced proudly that GM would be investing $600 million to upgrade the Fairfax plant in Kansas. A large chunk of the millions was being spent to completely replace the existing paint shop. Did this plant have an older paint shop? No, in fact the Fairfax plant was the last to be renovated under the Paint Plan and should have, by Godfrey's predictions, had technology that needed no further updates.

"Tricky Technology" is what the *Wall Street Journal* labeled GM's woes in May of 1986. The subtitle of the article "Auto Makers Discover 'Factory of the Future' Is Headache Just Now" tells it all. Describing the Hamtramck plant as more of a basket case then a showcase, the author goes on to note that "In the ultra-modern paint shop, robots at times have spray painted each other instead of the cars. Some cars have been painted so badly that GM had to ship them to a 57-year-old plant to be repainted."[24] Rushing into dependence on new technically advanced systems, not yet fully tested, is foolish at best. Volunteering and guaranteeing the use of underdeveloped systems in not one place but in many facilities by a predetermined deadline was crazy and nonsensical. But that is what GM did!

Maybe the failure of so many new plants could be written off as the cost of pushing technology too hard and could be forgiven as a noble GM gesture to reduce pollution; but even the EPA knew that environmental compassion was not what drove GM. It was the industry competition with Ford and Chrysler.

## Paint Inefficiencies in Application and Purchasing

While lack of foresight was creating unnecessary expenditures, the lack of oversight was contributing to the problem as well. As documented in an earlier chapter, there were numerous inefficiencies that went unchecked at GM, and undetected inefficiencies dogged the painting process.

It's a straightforward ratio. The more paint you spray, the more VOCs your paint booth emits and the higher your costs for paint. Therefore for purposes of emission reduction and cost control, it is ideal to have as high a percentage as possible of your paint land on the car body. Despite the high efficiency that was being obtained in the application of primer paint using electrodeposition (ELPO), the topcoat application process was far from perfected at the implementation of the Paint Plan and the overspray added to both the overall production cost and the volume of VOC emissions.

Almost everyone is familiar with the hand held spray gun, which uses compressed air to atomize the paint solvent mixture and force it out of the gun nozzle, hopefully toward the object to be painted. This, however, can be an inefficient process. For years, management was always told by GMAD that its paint spray application techniques produced an overall spray efficiency of 50%. In other words, if you spray two gallons of the paint at the car only one gallon actually adheres to it. Until engineering studies were required by the state of California, no one in management questioned the spray efficiency data or perhaps even thought to ask for it. Most of GM's spray efficiency numbers were mere assumptions. In 1971, with rising California regulations, engineers were forced to actually measure the spray efficiency at the Van Nuys, California, plant. It was found that the average spray efficiency was actually 25%, meaning that for every four gallons sprayed (or used), only one gallon of paint hit the vehicle. Thinking back, I wonder how losses like this missed management's attention for so many years with so many executives coming from the financial staff.

Paint efficiency continued to be an issue as the Paint Plan was implemented. When the new technology was up and running in the six new plants, no one in upper management realized that the efficiency of paint application was even worse than in the old plants until my presentation to the vice chairman of the board in 1986. The assembly plants built by GMAD and the Fisher Operational Group between 1983 and 1985 were touted to incorporate the latest in technological advances. They utilized computer-controlled robots and highly specialized mini-bell spray guns. The mini-bell spray gun has a spinning disc travelling at hundreds of revolutions per minute, sheering the paint into small particles. To enhance the guns efficiency it has a shroud enclosing the release area of the gun directing the paint to its target. Furthermore, the paint is given an electrical charge to cause it to be attracted to the vehicle surface and air jets give the paint an extra push. All these robotics accomplished the job of removing human workers from this toxic environment but amazingly, with all their design advances and good theory, the spray guns were found to be operating at a mere 15% rate, meaning they sprayed six to seven gallons of paint to get one gallon on the car! Shouldn't there have been some system in place to measure results and create accountability? Shouldn't comparisons have been made with other paint systems, and shouldn't that knowledge have been shared within the corporation?

If the VOC emissions caused by paint inefficiencies didn't garner concern, one would think that the rising paint costs would. A gallon of color topcoat in 1970s averaged \$6–\$10 per gallon and the thinner or solvent 60 cents per gallon. At that time, a barrel of oil was less than \$20. The cost of all coating materials to treat and paint a car was \$30 or less, not considering the overspray or waste, which could be two or three times the $30 figure. By the late '80s the cost of paint, which was controlled by the price of oil, was five times what it was in the 1970s. By 2011, a barrel of oil was over $100 and the cost of all paints had risen proportionately. Among them, the color pigments used in the topcoat were by far the most expensive.

Even our paint purchasing methods were not cost-effective. Our relationships with suppliers were not well thought through. GM contracted with paint suppliers to buy a predetermined amount of paint for all the plants from one supplier for the year or multiple years. The Japanese have a different system. They pay only for the paint that they actually use. Toyota buys its paint by the ounce. It weighs every container on load-cells before and after use, receives credit for any paint left in the bottom of the container and charges the vendor. In addition, their vendors must improve at least two or three colors each month to continue as a supplier. The GMAD paint engineers and their suppliers were in constant battles over development changes and who was responsible for what costs.

Before the Resource Recovery Act was enacted, a number of GM employees were fired for dealing in paint and paint solvents under the table. Those who weren't in bed with the suppliers were often duped by them. In the early 1970s, while touring Tarrytown's New York paint shop with Chevrolet's Regional Manager Max Mosher, we were stopped by Mel Halstead, the GMAD paint expert for Godfrey. Holding up a Vega door hinge that had undergone an accelerated salt spray test, he declared, "The vendor said his paint could be applied over grease but look at the rust!" Vega cars were having serious rust problems, but so were most of GM cars at the time. The regional manager nodded, as if to say, "Okay…fine." As we walked away he said, "Imagine—that's my paint expert."

## Squelching of Paint Development

So who exactly was a "paint expert" within the corporation? With all the talk of leapfrogging Ford and Chrysler, one would think that R&D in the paint department would be of the highest priority. In fact, as discussed in the Godfrey chapter, the exact opposite occurred. In-house research was stifled, records of abatement effectiveness were destroyed rather than compiled, and offers of paint advances by

PPG and Inmont Corp. went unanswered. I have in my files letters from these two corporations on their letterhead, both dated from the fall of 1978 as the Paint Plan was going into effect, indicating that they would be ready to supply GM with the lower emitting dispersion lacquer by the 1980 model changeover. These companies were ready to help GM meet the EPA standards. Why was there no follow-through?

In addition, no research on powder paint coating was allowed. Since powder coating is dry, all paint overspray can be captured and saved, reducing air pollution essentially to zero. As stated earlier, in the late 1960s I had Paint Research and Development Manager Gene Leithauser, working on powder coatings, but Godfrey had curtailed this research. His excuse in an internal memo dated August 3, 1979, was that "Powder continues to be limited as an automotive coating since it is not available in metallic glamour appearance and two-tone processing." But research should never have been placed under the director of an operating group. In order to develop true innovations, research should always be unhindered. Remember the base coat/clear coat system that GM eventually adopted from the Japanese? The clear coat is indeed a powder coating. GM could have been the leader instead of the follower if Godfrey's assumptions had not been followed.

There were many other episodes of upper management buying into the philosophy of holding back or controlling advances that would reduce air pollution created by the paint process, even when those advances could also reduce manufacturing costs or improve product quality. Up until the approval of the Paint Plan they were determined not to approve any advance that might set a lower pollution standard. An example of this involved a Cascade dishwasher additive to be used in waterborne paint. The additive was good for spotless water glasses and spotless paint. It was tried and tested by paint companies and GM engineers. The additive provided good results and showed promise in saving thousands of dollars each week at every plant using waterborne paint.

A meeting was held in the early '80s to discuss this additive. It was attended by James McDonald, the company president, Howard Kehrl, the vice chairman, two group vice presidents, one or two vice presidents, several corporate attorneys, thirty to forty engineers of all levels, and me. General Motors had terrible business problems back then and a meeting involving so many executives was sure to be humorous, if not sad.

After about a half hour of debate, I told the lawyer sitting next to me that the law prohibits companies from scheming to restrict environmental control technology. The lawyer did not say anything, so I stood up and very nervously said, "This technology cannot be stopped. The company selling this additive to GM is telling thousands of other companies about their product because it saves money and reduces pollution and waste. The vendor will inform our competitors and the environmental regulators of their product because they want to sell all they can, meaning its use cannot be stopped nor kept from the agencies."

Howard Kehrl then spoke: "We can't stick our head in the sand! We should use it, right?" Almost everyone nodded in agreement.

The meeting ended with this great decision-making and as I stood against a wall outside the corridor in the hallway, the company president, James McDonald, passed by me and said, "Nick, let's make it more exciting next time." I am sure he felt it was a waste of his time and that of many others.

Resistance to advances like this was facilitated by an atmosphere that limited communication within the corporation. There was a complete absence of shared knowledge on many topics, including that of air pollution reduction. Some of the blame for this could be placed at the feet of the historical structure of the corporation. Former GM employee John DeLorean details in the "tell all" book *On a Clear Day You Can See General Motors,* written with Patrick Wright, the tremendous effort it took to create paths of communication in the Chevrolet division alone.[25] But the employees that came out of Fisher Body, and Godfrey in particular, seemed to make an effort to block communication and the sharing of knowledge. We operated differently in the Environmental Activities Staff and I knew that communication could take place within GM.

In the programs for water pollution control and waste reduction there were few restrictions to communication and a free exchange of ideas flowed between plants, central offices, the Environmental Activity Staff, and even other companies. To this day, Water Pollution and Waste Management Associations conduct national and worldwide conferences to facilitate the exchange of new ideas and new technologies. But this was not true in the case of paint emission abatement.

For the purpose of reducing water and air pollution, GM engineers developed sulfur dioxide ($SO_2$) scrubbers, the side-stream collector, and plating system recycling techniques. They exchanged these and other developments with the world. Yet, not one new development related to *paint* emissions was allowed to be shared for fear it could establish a tighter standard.

There are hundreds of engineers throughout the corporation that spend millions of dollars on water pollution controls. Millions more are spent reducing waste and tracking every bundle of waste to its grave. And still, paint emissions and their control are kept close to the chest, in secret. The most startling fact that no one considers is that water pollution programs and waste management programs do not directly affect the sales of GM products, whereas the quality of the paint finish on the vehicle certainly can. Any smart business would concentrate resources and encourage collaboration on the pollution problem having the greatest impact on the bottom line, but not GM.

## No Internal Accountability

In 1983, six years into the Paint Plan, a number of the plants were burdened with financial troubles related to the Paint Plan. Four or five plants had not done anything in the six years to comply with the Plan; not even a line on a blueprint had been drawn. There were two new plants, Lake Orion and Wentzville, but instead of being problem free, both had inefficient paint systems and were oversized. There were legal challenges being issued against two plants, and millions of additional dollars being requested to correct mistakes. But even with these ongoing problems, management did not demand a full accounting.

I made several attempts to report concerns to upper management through the Environmental Review Board, but GMAD management,

D.D. Campbell, originally of Fisher Body, and his followers, did everything they could to withhold vital information. They wrote reviews attacking me, accusing me of not being "a company man nor a team player," and even requested my discharge more than once without success.

Remember LeFauve's "Don't Tell Roger"? I was with Group Vice President R.G. LeFauve and a corporate environmental attorney discussing with Charles R. Sharp his plant's state of non-compliance when LeFauve startled us with the revelation that no one was to tell Roger Smith how badly things had gone. He continued without hesitation: "Roger was promised the community odor complaints would go away. All the plants have problems. He was promised all the plants would be compliant with the regulations…they are all out of code. He was promised GM vehicles would have a world class finish…they are anything but that." LeFauve let me know that he was bound by the same rules and could not do anything to establish a new environmental standard nor install stack controls on a spray booth. Even a vice president appeared to be under the thumb of Godfrey's group and there was more than one person who paid the price.

Not one to be intimidated, I was emboldened by LeFauve's words so I decided to disclose the sad state of GM's affairs to the executive committee members. But before I could speak, GMAD questioned my company loyalty and went to the president, who turned the investigation of my cooperation—or lack of—over to the vice chairman of the board. Every time an effort against me was made with such force, I knew the right chords were touched. I continued in my job, but I could not get my concerns to be heard.

Years slipped by. It was now the early 1980s, the first of the new paint systems with all computerized technology in place had finally been tested for compliance and found to be only 18% efficient. Upper management first learned of this shortfall during my presentation to the Vice Chairman of the Board Howard Kehrl in 1985. Should they not have known sooner?

Just prior to that meeting in 1985, my vice president, Betsy Anker-Johnson, requested I be given a week to prepare a presentation and to answer the vice chairman's questions. But Kehrl gave a quick, sharp response: "No, Wednesday morning. 8:00 a.m."

This was a tight turn around, but I had been ready for years and did not need more time.

The vice chairman brought to the meeting Manufacturing Staff Vice President Alex Mair, the supervisor of Paint Development, and Gabe Tiberio, director of Manufacturing Development. To open the meeting, the vice chairman, with his finger outstretched towards me, demanded that I answer the following questions: *Who is thinking of the customer? Who is thinking of the paint superintendent? Who is thinking of the good of the company and the stockholder?* These were questions that should have been asked years ago and every day during the Paint Plan.

My reply was, "I am. Those are the questions to be addressed!"

I quickly recapped the entire Paint Plan history going back to Rule 66 in California. I explained the false assumptions regarding paint application efficiency at both the existing and new plants. I then went on to set before them critical objectives to which more attention should be given:

- Reduce paint waste: Everyone is concerned about paint waste disposal costs increasing from $3 to $30 per cubic yard. But nothing is said about the $2,000 to $3,000 per cubic yard of paint in waste. We can't control the cost, but we can control the amount of waste.
- Prioritize paint quality: The paint superintendent's primary objective has been narrowed to meeting a regulatory limit. The objective should be providing the customer with a world-class paint finish.
- Consider abatement options: Control devices should be developed and put into use to abate emission and thereby shift the focus from regulations control to quality control.
- Develop new technologies: A top priority should be the continuing development of new paint technologies. Examples on this point were provided with Vice President Alex Mair adding his support and ideas.

The vice chairman's original hostile attitude quickly changed, and there ensued a dynamic two-and-a-half hour discussion of the facts. The vice chairman thanked me and that afternoon his chief of staff called saying what a great job I did. "The vice chairman hasn't been as excited in the last two years as he is today. He's going to work on this situation. Thanks again." That made my day.

Soon after, the plants sent in appropriations to redo the robot spray stations. Paint application efficiencies increased, with some becoming over 80% efficient. It proved that if told the truth, management could help correct problems. But why did they wait so long to seek out the facts? Why did CEO Roger Smith, a top-notch financial auditor (some people granted him genius status), not order a continuous audit of the Plan's progress, especially after the conversation he had with Godfrey in 1976? Why were so many meetings under presidents Stempel and Reuss spent reviewing reports on the latest office moves with nary a mention of the auto production taking place in our plants? How can these failures be explained? The fact that top management was surprised to discover the follies of Godfrey's plan so late in the game, says a lot about the corporate environment during my years at General Motors. It was a corporation that majored in minor concerns and minored in the majors. So many problems with the Paint Plan were systemic.

## Fighting the EPA— The Cost of Resistance to Regulation

The corporate environment in regard to the natural environment was a final contributor to the financial failure of the Paint Plan. The entrenched attitude of regulatory resistance cost GM billions when it came to the Paint Plan. Consider again the files of letters that were written by GM engineers and legal counsel arguing that GM "had done enough" and should be left alone by the government. Something commonly heard about the company was the complaining about too much government in our business. Yet the Paint Plan invited the agencies, state and federal, to audit and enforce every plant's paint application techniques and the conversion schedule. It set GM up for potential fines that could have been avoided.

So much manpower and corporate resources were wasted fighting the inevitable progression toward vehicles that were safer for the passengers and safer for the environment.

The Paint Plan created serious legal problems for dozens of the plants. In fact, at no time during the 10-year program and later years was GM not in a legal battle with an environmental agency. It was a constant distraction from the real business of automotive assembly. Between the legal fees, court costs, and the EPA fines, these battles cost GM hundreds of millions of dollars. But there are two cases in particular that are worth highlighting.

## Framingham Loses in the Supreme Court

The Framingham compliance situation was not much different from the others with the exception that its case went all the way to the Supreme Court. First, let me recount several other facts that made this plant's story a favorite of mine.

My first trip to the plant is still as clear as if it were yesterday. Approaching the plant, on the left side of the road was a red brick wall about eight feet high covered with vines and beautiful flowers. This sight made me think it was the estate of some wealthy person. Farther down the road appeared a long high fence with razor blade rings at the top. Getting closer, a second fence about two feet from the first became clear. Right away I assumed it was a prison. But as it turned out, the razor wired fence was around the GM plant and the brick wall enclosed the state's prison for women.

Compliance problems for all assembly plants were primarily due to the problems discussed: the total lack of coordination with the product plan as well as the development lag in paint technology.

Remember these factors in reading the Framingham's compliance history:

- The original Paint Plan committed the plant to take an initial step of utilizing 27% dispersion lacquer topcoat paint by December 31, 1982, allowing five years for compliance.
- In 1982, not ready and not desiring to convert to 27% dispersion lacquer, the plant requested an extension to December 31, 1985, with the offer to go directly to a new paint with "waterborne equivalency" emissions by that date. When justified, the Clean Air Act allowed an extension and the GM offer was approved within 90 days of the request to the EPA. In the interest of public health, and to keep pollution control programs moving forward,

Congress had inserted the 90-day approval window to force the agency to make timely decisions.

- Again, having done nothing to comply, not even a line on an engineering drawing, on December 27, 1985 (four days from the date to be compliant), the company belatedly got the Governor (surely under pressure) to grant a *second extension* to August 1987.
- The Governor's extension was submitted to the EPA but with no response after 90 days, GM assumed the extension was approved.
- In August 1986, the EPA issued a compliance order alleging the plant to be noncompliant from January 1, 1986, to August 1987 with a potential maximum penalty of $17 million. The legal issues were: 1) Was GM entitled to the two-year extension? 2) Did the EPA's failure to decide in 90 days preclude their right to enforce the law?
- The case went to a Massachusetts trial court in May 1988 but the EPA's complaint was dismissed. The EPA appealed the dismissal to the First Circuit Court and disapproved the Governor's extension the same day. Oral arguments were expected to be held in January 1989. Not waiting, GM appealed EPA's denial of an extension to the court and filed an administrative petition on November 15, 1988.

A short time later, in early 1990, upon arriving to work, my director, David Eden, notified me that he had approved the company's submission of the Framingham case to the Supreme Court. I suppose that he thought that the court, with two recent Reagan appointees, would support industry over the EPA, but my immediate response was disbelief. "It's the worst possible case and you're going to lose. The plant has done absolutely nothing to comply in the last 10 years."

Upset, I called my friend John Calcagni at the EPA offices in Durham, North Carolina, and discovered "they were dancing in the halls." They couldn't believe GM was so arrogant.

The case was finally argued before the U.S. Supreme Court on March 21, 1990.[26] The day of the Supreme Court hearings, the room was overflowing with industrial representatives and lobbyists because their decision would have a major effect on not only the automobile industry, but also the steel, chemical, energy, and other manufacturing companies. Three months later, Justice Blackmun delivered the unanimous decision of the court.[27] The industry lost nine-to-zero, with the case making it clear that a 90-day time frame does not apply to any second extension request but only to the first extension, which in this case was made back in the '70s. The court also noted that GM did not even attempt to reach compliance and should have known better. The case set a precedent for all industries.

This legal case could have been avoided. After years of delay and an extension, with no hard evidence of an attempt at compliance, the judge's decree essentially said, "What was GM thinking of getting away with?" Clearly these cases should not have been pursued or supported.

What is really sad about Framingham's case is that their paint system was relatively new at that time. I saw it built in the late '60s or early '70s, and yet GMAD told top management they wanted a completely new shop. Why? Throughout GM there was a practice by the individual plants to request new paint facilities, more power house boilers, and additional plant floor space. The idea was that, if approved, there would be less likelihood of the corporation ever closing their plant. Seldom, if ever—and I was in central office construction management for six years—did a plant fully utilize their existing systems. Powerhouses, air compressors, and electrical substations were all overbuilt by a factor of three or four over needed capacity. The thinking was that since a new paint shop or powerhouse represented a large capital investment, the plant would be protected from closure and would surely be selected over an older plant for the production of a new product. This was logical but not in line with reality.

At Framingham, a completely new paint shop was finally constructed, but the plant was shuttered a couple of years later. The plant in Van Nuys installed Framingham's new paint rotational guns at great expense, and then it too was closed. The building of new facilities, the upgrading of old ones and the subsequent closing of the plants became common and cost GM billions.

Framingham was only one example. There were serious compliance problems at most of the plants.

## GMAD—LORDSTOWN, OHIO

The GMAD assembly plant in Lordstown, Ohio, also committed to a 1985 compliance deadline as part of the 10-year Paint Plan. Remember that the yellow book stated that GM needed time to develop plans, make engineer drawings, and order equipment. After seven years of doing nothing, GM requested, again at the eleventh hour, that the Ohio Governor grant an extension. Under the pressure of the corporation's threat to close the plant, the governor agreed and signed the extension.

Ninety days passed and the EPA had neither approved nor denied Ohio's extension request, so GM again assumed that the extension was accepted. Later, the EPA wrote GM stating that they were in violation. GM still did nothing, counting on the extension signed by the governor. The Legal Staff supported the Lordstown assembly plant and tried to justify GMAD management's failure to do anything to comply with the law right up to the judge's decision. This time the EPA initiated the court case and again they won. According to the EPA "Enforcement Accomplishments Report - FY 1991" (April 1992) in the case of *U.S. v. General Motors Corp.*, (N.D. OH), "A consent decree was entered on April 11, 1991, resolving the government's action under the Clean Air Act (CAA) against General Motors Corp. (GM) for violations at its Lordstown, Ohio, automobile assembly plant. As a result, GM installed a coating system that reduced the volatile organic compounds (VOC) [coming] from its paint shop operations from approximately 3,400 tons per year to 750–800 tons per year. GM paid a civil penalty of $1,539,326."[28]

This fine of over $1.5 million was the highest in industry history and it could have been easily avoided. *The following Monday, after the court decision, the Lordstown plant was in compliance* with the use of a high solid lacquer paint. This paint change immediately reduced the VOC emissions as indicated in the EPA report and GMAD plant engineer Bob Phillips reported that the change actually saved GM money in overall paint cost.

Back on August 3, 1979, Godfrey stated in an internal memo, "We started development of a higher solids lacquer (17% LDL) in Lordstown with DuPont and have continued this development..." Why was the development not completed and acted upon during the next 12 years? Consider what this case tells about the cost of corporate mismanagement and of GM's policy to resist regulation and fight rather than find and apply solutions.

The entire Paint Plan and all its costly pitfalls could have been avoided with a simple conversion to high solids lacquer. The money should have gone to the research engineers instead of the lawyers. Management spent time avoiding doing the right thing instead of concentrating on the product success. The prideful attitude of noncompliance was a major contributor to the failure and expense of the Paint Plan.

## Relief from Walt Barber for Pontiac Truck

Again and again, I saw situations where GM could have saved millions of dollars through honesty with the EPA instead of resistance and deceit. Back in the late '70s, Vice President David Potter and I were invited to GMAD to discuss the Fremont, California, noncompliance issue. I informed Potter that Godfrey had been granted an appropriation of $33 million to install the carbon adsorption system and bring the plant into compliance, yet Godfrey was going to bluff to the EPA and tell them that the final cost would be $600 million.

Potter quickly replied, "If Godfrey tells the EPA $600 million and only spends $33 million, I'll make him burn the difference in the parking lot!" But it was a lot of bravado and I knew it. No one stopped Godfrey from lying to the EPA. He leveraged the false figure to enter into a court decree to buy time and then build a carbon system that was only operated for a year. The $33 million could probably been saved if the EPA had been informed of the fact that the Fremont plant was being considered for closure.

My position afforded me regular contact with the regulators of the EPA. I learned from experience that there was a better way. During my involvement in a Pontiac truck project, I met an environmental regulator that I came to admire: Walter C. Barber. He was the Deputy Assistant EPA Administrator and temporary head of the Federal EPA under Reagan. His approval of the Pontiac truck project without applying the justifiably more stringent control levels encouraged me to always present GM cases truthfully to the agencies. Barber could appraise a difficult situation and evaluate people as to their character, talents, and ethics better than anyone. I sought his advice throughout my environmental career.

Many engineers believed that all environmental groups and state and federal agencies were generally opposed to industry and would not grant them relief from the law, even if it were a reasonable request. The feeling of opposition manifested itself in years of fighting tooth and nail against automotive and safety regulations, most of which became laws to protect public health and safety. Some companies had engineers who did not conceal their hate for the agencies, and always failed to present their cases wisely.

From the perspective of an environmentalist, I found the management in the Michigan Department of Natural Resources, the Federal EPA, and the major environmental groups to be extremely intelligent and reasonable. Cases presented to them that were just, truthful, and proved no serious harm to the environment were always approved. Examples of the fair practices of the non-profit Natural Resources Defense Council (NRDC) are numerous. General Motors alone saved $18 million in $CO_2$ emission controls because of support from Dave Hawkins of the NRDC.

By 1973, GM had missed most of the small-boxed pickup truck market. Possibly two or three years of sales remained, so a proposal was made to install a small assembly line with paint operations in an old vacant plant in Pontiac, Michigan. The area was designated as non-attainment for ozone, meaning the LAER (Lowest Available Emission Rate) had to be applied, the cost of which jeopardized the project. The production rate of 6 to 8 vehicles per hour with a possible maximum of 18 per hour was planned. This amounted to about one-fifth of the normal assembly plant rate of 70 per hour. The three key reasons given for an exemption from NSPS were: Pontiac's unemployment rate was extremely high, the plant would be in production only a few years, and enamel paints, not lacquer, would be applied, making little measurable difference in the air quality.

The Michigan DNR could not approve the exemptions, so I took the case to Durham, North Carolina, and presented it to Walter C. Barber, deputy assistant administrator for the Air Quality Planning and Standards Division of the EPA. Before I got off the plane that day in Detroit, Mr. Barber had cleared the exemption with the Chicago EPA regional office and the state DNR.

As expected, the plant operated only two years because of the poor sales and finally closed, never exceeding 10 trucks per hour. The cooperation between the EPA and GM benefitted both parties. In the same way, could GM have worked cooperatively with the EPA to obtain a separate RACT to accommodate continued use of lacquer and thereby avoided the commitment of the Paint Plan? I am certain of it.

## The Truth Pays for Jan Tannehill

Even with the Paint Plan in place, GM could have gotten relief if they had worked with the EPA as soon as they knew that the plants could not comply. The company did get relief for six plants and for Saturn by telling the truth about the company's problems.

Jan Tannehill, BOC (Buick, Oldsmobile and Cadillac) regional manager over six assembly plants, asked me for assistance with an EPA matter. I took him to meet the EPA administrator for Air Quality in Washington DC. In private, I told Tannehill to tell the EPA administrator, J.C. Potter, the truth regarding our poor products, poor manufacturing systems, new technology that was not working, and what decisions management made that turned out badly. I told him to be as truthful as possible. I would not allow anyone from Industrial Government Relations, Legal Staff, or Central Office to be in the office with Tannehill and the administrator. On the plane back to Detroit, Tannehill said he told the administrator everything—whatever that meant.

That same night I got a call from John Calcagni, in the EPA Durham, North Carolina, office with surprising news. He said the EPA administrator called his staff into his office and told them that Tannehill was the first industry representative to tell the truth—that they needed help—and said, "Do everything you can for him." Not one of his plants was audited or ever received a violation notice like Framingham and Lordstown. Our friendship lasted into my retirement.

Looking back through the lens of time, it is easy to see that the Paint Plan met none of its promises. It failed to produce a world-class finish, it did not solve all pollution regulation issues, it did not cause GM to leapfrog past Ford and Chrysler and, as I tried to tell management from the beginning, it was not the only option. So what did I do at the conclusion of the 10 years? I threw a party.

# 14

# The Costs Continue

The 10 years of implementing the Paint Plan were hard on everyone and, looking back, the most chaotic in the company's history. A year after the Plan had ended, the GMAD Central Office was still in a panic trying to improve the systems in the new paint shops, and six new plants had compliance problems. In spite of the setbacks, Roger Smith was still claiming the assembly plants would be the producer of world-class products by the mid-1990s.

My relationship with GMAD's operating staff was as strained as ever as they circulated a report claiming I was not a team player and should be removed. But in 1988, at the end of the 10-year Plan, I was an assistant director of Corporate Air Quality Control and enjoyed more support from my management than ever before. In fact, strange as it may seem, I look back on those Paint Plan years as my best, most enjoyable years. Why? Because I have never forgotten my roots. Even the opposition was, to me, a sign that my opinion had become respected and even feared. Who would have thought that the words of a man who began as a trade school graduate could be a threat to top executives at one of the world's largest corporations? It amazes me.

So even though the Plan had been flawed and the implementation had been fumbled, I wanted recognition for my fellow employees in the trenches that had labored to make the Plan work. Compliance problems continued, but the 10-year monkey was finally off our backs, so why not show our class and throw a party? I presented the idea to my director, David Eden, who immediately took it to Vice President Betsy Ancker-Johnson with his orders to do it in a first class way. The proposal was to invite everyone now associated with the Plan, along with their spouses, to a dinner at the exclusive

Grosse Pointe Yacht Club in Grosse Pointe Shores, a wealthy suburb of Detroit with pristine lakeshore views.

The October 26, 1988, celebration was a great success with Walt Barber, former EPA administrator; James McDonald, retired GM president; and Robert Stempel, current GM president, taking part in the presentations on the Plan's history. I still have the invitation sent to Joseph Godfrey, the Plan's originator, at his home in Florida, but returned for insufficient postage.

I was toastmaster and made the final presentation with a surprise special recognition of Dave Aldorfer's efforts through the years to keep our lines of communication to GMAD open at some risk to his career. McDonald thought that was a great thing to do and told me, "That's the kind of party GM should have."

It wasn't until my retirement years that I realized the importance of the untold story of the Paint Plan. Over the four years of researching, writing, rewriting, contacting engineers and agency personnel, and verifying the status of the assembly plants via the internet, my resolve to share this story has only strengthened. Now

*Celebrating the completion of the Paint Plan, October 26, 1988, at the Grosse Pointe Yacht Club (L to R) Nicholas Kachman with retired GM President F. James McDonald and the Vice President of the Environmental Activities Staff, Dr. Betsy Ancker-Johnson.*

Credit: Unknown party attendee

I wrap up the final chapters with a burst of energy because of a recent announcement about the Fairfax, Kansas, assembly plant, which proves I was right about the inefficiencies of the plants built under Godfrey and the continuing drain on GM from the Paint Plan. I have no regrets about throwing the party, but it would have been less of a celebration had we realized then the magnitude of the Paint Plan's contribution to the eventual bankruptcy of General Motors. The related costs continued to mount long after that date in October 1988. The total cost of the paint program will never be known because there are so many ancillary costs associated with the Plan. But over the past thirty years, it is fair to assume that the figure is well over $100 billion.

Included in that total would be what I will label as the "5 Cs," namely, conversion, construction, corrections, "court" costs (legal fees and legal penalties), and closures. At each of these stages, there was loss of production and loss of income, or conversely, outlays of capital expense that would never be recouped. Follow the money.

## CONVERSION

Conversion of the paint shops was an expense anticipated at the outset. Everyone understood that compliance with the environmental regulations would be expensive, but the cost estimates approved with the Plan were not realistic. As stated before, the original Paint Plan covered 39 total paint shops, 33 of which were in 31 assembly plants, with 6 in Fisher Body plants. The original estimated cost of conversion was only $4 billion. Most plants converted initially to a lacquer that contained 27% solids or to enamel that contained 50% solids. If it were not for the promises made in the Paint Plan, the conversion process could have stopped there. The higher standard was, by law, only applicable to new construction, not to the modification of existing facilities. This step in itself was modest because it involved a change in paint formula but not in spray equipment, yet it was taken at a cost of tens of millions at each plant and that was only the beginning.

Twenty plus plants then converted from lacquer to color topcoat waterborne paint or had new paint shops constructed. The cost per plant ranged from $25 million to over $600 million at each plant—the total? Tens of billions. For existing assembly plants to go beyond reasonable reductions of emissions at a reasonable cost and switch to waterborne topcoat paint required tens of billions of construction plant changes for a rather small amount of reduction of the paint emissions. It cost billions, yet created no increase in sales. Instead of leapfrogging the competitors, it caused GM to fall behind in profitability. The actual cost of conversion went well beyond what was approved as well as what was necessary and every penny of that was lost when within a few years many of those plants were shuttered.

## Construction

The original Paint Plan chart only provides dates for the *conversion* of existing plants, but using the justification of the Clean Air Act coupled with the switch to platform assembly, *eight* new assembly plants were constructed during the 10 years of the Paint Plan. It was an unheard of pace for construction. The shift to front-wheel drive, also occurring during this time, did change the assembly configuration and required more headspace over the assembly line, but not always an entirely new plant. Many of the construction projects found justification in the requirements of the Paint Plan as well as the corporate attitude that times were changing and everything must be newer, grander, and more modern. In my opinion, it was the plan for new paint shops that drove the frenzy of construction.

Never mentioned by any of the Paint Plan's yellow book writers were the potential costs associated with construction of new plants and the cost of closing and performing environmental cleanup at the old plants. These additional expenses represent billions beyond the $4 billion Plan estimate.

Over the first 50 years of GM's existence, assembly plants were always added in response to marketing success. They were only closed when they became too old to regain operating efficiency. With few cars produced for the public during World War II, plants that produced military planes and trucks during the war were quickly renovated afterwards to meet the pent-up demand for civilian vehicles.

Two years after the war ended, in 1947, the Framingham, Massachusetts, plant was built to supply the increasing East Coast market. Then in 1952, five years later, the aging Oakland City,

California, plant was replaced by the new Fremont plant not far from Oakland. Fourteen years later, in 1966, the Chevrolet Division sales were outstripping those of all the other divisions combined, and thus to increase Chevy production capacity a plant was constructed in Lordstown, Ohio. With increments of 5 and 14 years between openings, these plants were added at a pace at which there was time for analysis and improvement of one plant before another was built. This matched the norm in the automotive industry.

In 1973, Godfrey, recently given control of both divisions, ordered the construction of a new assembly plant, the first in seven years, at Oklahoma City. With delays caused by the oil embargo, the plant didn't open until 1979. As discussed in the Godfrey chapter, it was a plant that was outdated long before its doors opened.

Construction of any assembly plant involves a sizeable outlay of capital. Given that fact, one would think that industry "best practices" would be utilized at every step, but not at GM. Ignored in the application of the Paint Plan were the basic engineering approaches taught to all engineers for the development of production and manufacturing systems. The principal protocol is to build one system to a design criteria using engineering facts and assumptions, measure the result, compare the result to the factors used, discover errors, correct the assumptions, rebuild until the goals are met, and then—and only then—build more systems. In the '80s at GM, this did not happen. Following on the heels of the 1979 fiasco of Godfrey's Oklahoma plant, GM built three oversized plants and converted other plants, applying unproven technology, in order to meet the compliance dates promised. This accelerated construction left no time to test actual performance and compare it to the original design criteria. All the plants failed to comply. Millions of dollars were requested to redo all the plants. Eventually, it took innumerable modifications over the years to reach consistent functionality and to obtain reasonable spray efficiency rates. These continuing modifications came at a very high cost.

It was assumed that the GM assembly plants, using all of the latest advanced technologies, would be the most productive and modern in the world. Remember that the goal was to have GM known as the world-class producer of the most glamorous, high-quality, low-cost vehicles in the world. There seemed to be no restraints on funding. The bigger the plant the better, the newer the device the better, the more robots the better, the more computers and computerized programs the better, the more stainless steel over plain steel the better, and so on. The new plants were to be the best in all categories.

So when the massive undertaking of modernizing the North American assembly plants got the board's approval, construction of three brand new plants started with the Lake Orion, Michigan, plant (1983), Wentzville, Missouri, plant (also 1983), followed by the Detroit/Hamtramck plant ("Poletown," 1985). Amazingly, three assembly plants were under construction off the same drawing at the same time without the performance of the first having been verified to the design criteria. I have to wonder if a written design criterion even existed.

During the same period, three new truck plants were also constructed—in Shreveport, Louisiana (1981), Fairfax Assembly in Kansas City, Kansas (1985), and Ft. Wayne, Indiana (1986). It was a boggling outlay of capital with so little accountability.

The opportunity to do great things was placed in the hands of Leonard R. "Len" Hostetter, director of Operations and Engineering for both the CPC and BOC divisions. Len Hostetter is best remembered for modernizing Fisher Body and GMAD's press operations, and reducing the time to change press dies in the forming of parts from days to minutes. His was a well-managed program.

But a temptation to which builders sometimes fall prey is the tendency to build monuments by which to be remembered. He was a good engineer, an honest man, and had only the best intentions in the building of the new plants, but the temptation to build big was there. Described earlier as examples of an inefficient pollution system, the three new auto plants ended up with more than twice the floor space of the competitor's facilities. The Detroit Cadillac plant and the Lake Orion plant were so massive that four different large vehicles (two Cadillacs, one Pontiac, and a Buick) were produced at the same time while still leaving huge sections of the plant completely empty. A maintenance worker sent to do a repair job might not be seen for days! At the same time, the Japanese were keeping production costs low by building smaller, more efficient plants all under one million square feet.

The new assembly plants are listed below in order of their opening, but remember that all 39 Paint Plan plants were simultaneously attempting to comply with the law by converting their paint systems. Not to be forgotten moreover, is the company's market share which was losing 2–5% each year, requiring a great infusion of investment into the business or product plan. Yes, there was a great deal of chaos throughout the company during the Paint Plan years.

The New Plants

1979 Oklahoma City, Oklahoma
1980 Bowling Green, Kentucky (Corvette)
1981 Shreveport, Louisiana
1983 Orion, Michigan
1983 Wentzville, Missouri
1985 (Detroit) Hamtramck, Michigan
1986 Fort Wayne, Indiana
1987 Fairfax, Kansas City, Kansas
1990 Springhill, Tennessee (Saturn)

Nine new plants in just twelve years!

Following the construction of the Saturn plant, it was not until 2001, 11 years later, with the Lansing, Michigan, Grand River plant, that another assembly plant would be built. This was a more reasonable time frame. The condensed construction of so many plants in the rush to comply with the Paint Plan caused the company to waste millions needlessly in many areas, and particularly on the exhaust system that was designed as a way to avoid the use of abatement.

Had a basic study of electrical usage at Shreveport, Orion, Wentzville, Hamtramck, or Fairfax been performed and reviewed, the data would have revealed the enormous waste of power, the difficulties of maintenance, and the inflexibility of the design. A better design would have been found. That crazy design is still costing GM.

## CORRECTIONS

The promises of Roger Smith and Joseph Godfrey that the Paint Plan was a "once and done investment" should have been rebuked at the outset. Keller notes that by the end of the 1980s, GM President

McDonald was thoroughly disgusted with Roger's never-ending promises. In addressing graduates at Michigan Tech University he said, "Don't listen too closely to anyone who says he can tell you what the future is going to be like…We have no illusion about how much of an edge advanced technology will give us. It could be only a year or two before it becomes outdated by yet another generation of innovation."[29] True words finally spoken!

Every model change in an assembly plant had to include in its budget money to correct problems in the paint shop. Every year a portion of the plants with the new waterborne robotic operation required millions of additional dollars allocated to spray booth improvements. Combined cost: Billions!

The Paint Plan writers over-concentrated on the colored topcoat waterborne paint to meet the regulatory commitment. The yellow book writers were without vision of what was to become a continuous improvement in paint coatings, color changes, robotics, spray applicators, computer programs, and the need to transport the vehicle body through the spray booth with great accuracy in relationship to the robotic spray guns. The vehicle bodies wobbled from their overhead conveyors and had to be moved with floor dollies.

Thirty years later, advances in paint compounds and application technology continue. The most recent advances allow the colored base coat to be applied on the primer surface while it is still wet. The clear coat can also be applied on the colored base coat before it is dried. After the clear coat is applied, the vehicle travels through a drying oven for the final curing of the entire paint coating. This "wet on wet" process eliminates two out of four long drying ovens, accelerates the coating process, reduces energy demands, and by shortening the travel distance, greatly reduces the chance of unwanted specks of dirt landing on the paint surfaces. This new painting process will lower construction costs and energy usage while providing improved quality control.

Articles from 2013 regarding a brand new paint shop at the Fairfax, Kansas, plant, which was originally built in 1987, boast that the energy savings from use of the new paint shop will be 50%. My industry experience and engineering knowledge cause me to question this figure. Although no one will publicly admit it, I am certain that this savings is a commentary not just on the efficiency of the new "wet on wet" system, but also on the inefficiency of the original single stack exhaust system. There is no other way to account for the dramatic energy savings touted in the press releases when the physical structure remains comparable in size. I don't doubt that there is tremendous relief that the new shop will allow that plant to abandon the outrageously inefficient system for dealing with paint booth emissions that was left from the original construction. Just turning off the giant exhaust fans would result in significant savings. Time will tell if the three remaining plants with the single stack system will continue to operate under the expense of this system, be shut down, or be allocated the $500 million that it takes to completely replace a paint shop.

Within a few years, the promise of "once and done" proved to be either disingenuous or naive. With corrections to almost every step in the painting process from procurement to fume emissions, a lesson learned from the Paint Plan experience is that the readiness of new technology cannot be predetermined and it is foolhardy to promise its use at an arbitrary future date. Paint suppliers will continue to pursue innovation in order to maintain or increase their market share. The 10-year Paint Plan appears more ridiculous as time passes. The promises it made went unfulfilled and this had legal consequences.

## COURT AND LEGAL COSTS

As well documented in the previous chapter, GM incurred significant Paint Plan costs associated with legal fines and the battle to avoid those fines. The fines could have been better avoided by not making promises that couldn't be kept, actively working to meet promises made, and breaking free of the knee jerk GM habit of resistance instead of cooperation. It never ceased to amaze me that so many man-hours were put in by engineers and lawyers writing letters to protest compliance demands. Why would a car company keep spending money, which only served to exacerbate the problem and delay solutions, instead of on solving problems and improving our relationship with both the government and the public? Millions could have been saved by keeping our focus on improving our product and its production.

The Environmental Activities Staff was equally responsible for handling issues related to water pollution, yet without the legal entanglement of the Paint Plan. Water pollution problems were handled proactively, promptly, and generally with cooperation towards the government. The legal costs were kept to a minimum. The orders to fight air pollution regulation rather than cooperate came from outside my department. We were left to shake our heads at the nonsensical priorities. I had sounded repeated warnings, but it was not until after the $1.5 million fine at Lordstown that consideration of future legal battles were submitted to the EAS, and based on my track record, they came first to my desk.

## CLOSURES

This final cost category is the most overlooked. The tremendous outlays of capital associated with paint shop conversions and plant construction are obvious. The ongoing costs of corrections and legal fines are a well-documented reality, but few consider the losses that come with plant closures beyond the employee layoffs. Those layoffs deal a harsh blow to communities that had relied on the assembly plants for jobs and taxes, and all the related services, but there are substantial monetary costs to the corporation as well.

Every plant closure represents a significant loss of capital investment. The more recent the plant's construction, the greater the loss, since no return will be forthcoming on that investment. With that thinking in mind, the management of many GM plants assumed that recent capital improvements would save them from "the chopping block," but that was not enough during and following the Roger Smith years of rapid growth.

Also to be counted, in addition to the lost return on investment, are the outlays of cash required to shut down a plant site. Under tightening environmental restrictions, any damage or changes to the plant's physical environment must be mitigated and reversed. Host communities often write into their zoning approval closing restrictions: any tax breaks must be refunded and the tab for bringing in utility and sewer hook ups, if originally proffered by the locality, now must be repaid. While it costs money to operate a plant, it also costs money to shut it down. Abandoning the billions in facility investments that were made in conjunction with the 10-year Paint Plan had to be a painful decision on several levels:

- Close to half of the original 39 plants on the "Conversion Plan" chart were targeted for closure prior to the end of the 10-year period. In the 12 years that 9 new plants were built, 24 existing GM plants were closed.
- Eight plants on the original conversion chart were shuttered by 1986. The remainder had new paint shops built or made major paint shop changes, but 10 more were closed by 1990 and another 10 were shuttered by the end of 2009. Today only 3 of the 39 are still in use. Most could have operated using 27% lacquer paint until their closure, saving billions of dollars in both direct and indirect costs.
- Two of the newer assembly plants are also out of operation and a third is operated on a very limited basis. Oklahoma City, built in 1979, was never efficient and was finally closed. Shreveport, built in 1981, is today an empty site. Saturn, built in 1990, and costing GM billions was improved in 2008 with an additional infusion of $330 million for the paint shop, yet today sits largely idle with future plans for only limited production. But the total cost? Tens of billions!

As recently as February 24, 2013, it was reported to me by several insiders that Godfrey's orders to avoid abatement controls on any of his spray booths, no matter what the additional cost of construction or operation, were enforced as soon as he took command of GMAD and Fisher Body. Godfrey's 1975 Oklahoma City plant and the 1981 Shreveport plant were the first to be closed, likely because of their high-energy cost, unreliable operation, and system maintenance problems. It has been nearly impossible to get any of the current management or engineers to discuss the underground designs or their energy usage.

Godfrey harmed the company more than any other executive, but his management team and the engineers involved in the Paint Plan deserve blame also. This is an example of the GM culture in control of the salaried workforce. The lowest cost assembly plant built in modern times is the Lordstown, Ohio, plant, built back in 1968, and was a project which I worked on as a mechanical engineer. It is one of the remaining three plants listed in the Paint Plan that is still in full operation today. The others were first placed in operation even earlier: Arlington, Texas (1953), and Flint, Michigan, Truck 1 & 2 (1947).

## Summary of the Costs—Waste

The total cost of the Paint Plan will never be known because there are so many ancillary expenses associated with it, but over the past 30 years, it is fair to assume the figure is well over $100 billion.

Included in that total would be the "5 Cs" that have been examined in turn. At each of these stages there was loss of production, loss of profits, and labor problems. Not considered in the cost of the Paint Plan is the indirect effect it had on product programs and sales. All of those dollars spent netted no, repeat, *no* increase in sales! These same dollars could have been spent on product improvement—product improvement that would have resulted in increased sales. Whenever possible, dollars spent should increase dollars earned. Without the Paint Plan, GM management could have concentrated on improving product plans and having the resources to enhance vehicle offerings. In the 1920s and '30s GM was a leader in innovation, but by the '80s the Japanese were at the head of the pack and GM was a mere follower. GM had been the pioneer for automatic transmission, but now all the transmission innovations come out of Japan.

In April of 1992, Stempel and Reuss, the heads of the two major car divisions, were fired, ostensibly because of the company's poor product showing and falling market share under their leadership. The Paint Plan has never been mentioned as a cause for GM's bankruptcy, but it had a huge impact. Never before exposed is the continuing drag that the Paint Plan had on GM's financial state and ability to fund product innovation. All the press knew was that the GM North American operations were hemorrhaging cash and that in light of this, heads were rolling.

Hours before Stempel gave what was to be his final speech as a GM executive, I was among the executives that management's top consultant called, in his attempt to advise Stempel. Hired by Roger Smith to give training seminars, the Boston-based consultant, Marc

Sarkady, was enlisting support to get Stempel to avoid saying yet again that "GM has turned the corner." I advised that Stempel admit that GM had problems and that tough decisions would have to be made, but announce that solutions were available that would allow the company to come out stronger in the future. I hung up the phone hopeful and pleased that my opinion had been requested. Later that day, just hours after his third "we've turned the corner" speech, Stempel was on his way out.

# 15

# What Must Change

As I stated at the outset, I have chosen to put my observations in writing because I think they tell an amazing story and because, in spite of its shortcomings, I still care about General Motors and want to see it succeed and prosper once again. When I first began to commit my story and my concerns into writing, I had lingering doubts about the validity of my viewpoint and my wife said I would lose my pension for airing my concerns. But over the five years of writing and research, my confidence and resolve have only strengthened. The facts support my hunches and the clear view of hindsight corroborates my concerns.

People have had different reactions as I've shared this book. A retired executive of Saginaw Steering Gear Division read several sections and declared, "What an indictment of management—I experienced some of the same things in my plant." But current employees won't touch it or even comment on it. My own brother, as he lay dying of cancer in a Los Angeles hospital, coincidentally next to retired GM Vice Chairman Howard Kerhl, was afraid to have his name associated with the book for fear that "they" will come after the family. Yet many others have made the same, if not similar observations of the flaws in GM—they simply were unaware of the connections with the Paint Plan.

Almost all of the lessons to be gleaned from GM's downfall could benefit the bottom line at any manufacturing business and many have broader implications for any organization. The financial crisis at GM has not generated enough serious public discussion nor, to my knowledge, have business schools made GM's failure a major topic. It should not be forgotten that what was once the biggest most profitable company in the world went bankrupt. Blame labor, health care costs, too many regulations, but never management for how they managed these problems. It was the company's culture, the daily life of the business that destroyed the once proud GM. The concerns that I have raised do indeed date back to the late '70s and the '80s, and in recent years GM has made progress, but as I have shown, only *some* of the core causes have been addressed. Many are still holding GM back from reaching its true potential. The seeds of GM's bankruptcy were planted in that era. In order for GM to thrive, the lessons of the past need to be studied and applied.

A GM spokesperson recently claimed that the culture at GM is undergoing change with the recognition of younger executives, the holding of smaller meetings and further easing of the dress code, but I can attest that these changes are neither new nor significant. They took place 20 years ago without ever touching the root of the problem. Significant change can only come under a leader with the courage and the support to acknowledge and openly confront the self-destructive power of pride and self-interest, redirecting the corporate focus and enforcing the new priorities.

I was fascinated to read how a new president took the helm at Alcoa, the world's third largest producer of aluminum, and by radically redirecting the corporate focus and enforcing the new "habits," firing those who defied them, was able to effect a change in Alcoa's culture.[30] The result was a complete corporate turnaround. The culture at GM has been entrenched for so long that the few bold ones like George Eads and Ross Perot, who tried to change it, were forced to resign themselves to defeat. A few others lived under the delusion that their restructuring was successful in changing the culture, but time has revealed otherwise.

What I believe GM needs is a set of values that every level of employee can embrace. They don't need a lengthy list of financial goals like those that Roger Smith put forth at his leadership retreats. They need a rallying cry that will direct their everyday thoughts and energies. I have some specific recommendations to make but they all come back to making a change in what is valued. GM needs to value its customer. To do that, it needs to value research, problem solving, and the thorough analysis of cost/benefit, with constant improvements to both the product and the process. It needs to value the corporate resources already in its possession—both human and material. Gone are the days of resting on its proud history and the status quo, assuming that the customer will always be back. Gone are the days of seemingly endless corporate financial resources, casually tossing aside the "old" and constantly replacing it with whatever is "new." In a nutshell, GM needs to stop resting and resisting and it needs to begin reaching. It needs to stop resting on its laurels and resisting outside advice and all government regulation, and it needs to begin reaching out, reaching back, reaching down and reaching forward.

## Reach Out

There are three specific areas where GM could benefit from an increased effort to reach out. All businesses need to reach out to their customer base. Most can benefit from reaching out to the communities in which they are located. Manufacturers, in particular, need to reach out in cooperation with the government agencies that regulate them in a spirit of cooperation rather than antagonism.

**Listen to the customer.**

GM needs, first, to reach out to its customer base—both present and future. This seems like an obvious statement for any business but as I stated earlier, it was my observation in the '80s and '90s that corporate pride caused the opinion of the customer to be undervalued. There was no intentional and sustained effort to collect, much less respond to customer desires and feedback. In the bubble of Detroit, consumers continued to buy vehicles from the "Big Three," but across the rest of the U.S. customers were flocking to the Japanese offerings because they had fewer problems and lasted longer. Detroit automakers were still cutting corners to save money while the customers were willing to pay more for higher quality.

**Care about communities.**

Every corporation needs to have staff that present the public interest in meetings and memos and cause the business to ask how it can serve the wants and needs of the customer and the larger community, not only in product offering, but also in the areas of safety, quality, and environmental responsibility.

This is an area where there appears to be progress. I applaud the changes to embrace safety and environmental concerns. But particularly in the area of environmental issues, I have to question if the company is weighing the cost/benefit of where dollars are being spent.

In the last 10 years, GM has touted its emphasis on "Lean Manufacturing" and the "We Care" program—an acronym for "Waste Elimination and Cost Awareness Reward Everyone." They have boasted of becoming "Landfill Free" and in 2013 added their name to a list of companies that are "calling for federal action to confront climate change" by signing a "climate declaration," but behind the press releases and propaganda, the actual impact is shallow at best. It may boost their public image, but a closer look at the details by those in the business reveals more hype than reality. My peers recall that the manufacturing changes we made in the 1980s resulted in pollution reductions of upwards of 80%. The reductions that GM now boasts about are but 1–2% of the already greatly reduced amount. Numbers out of context can sound more impressive than they really are.

I recently read an article in which GM boasts that their factories are now "landfill free" but what does this really mean and is this goal worthy of the effort and expense thrown at it? While increased recycling and reduction of waste are worthy targets, there is nothing wrong with sending decomposable materials to a landfill. If an auto manufacturer wants to make a significant environmental impact, directing their resources at continuing to improve fuel economy would be more meaningful. A one gallon per mile fleet average increase in fuel economy by GM would make anything Lean Manufacturing provides insignificant.

Every company is wise to consider if there are ways that it, as a corporation, can serve the greater good of humanity. Are there environmental projects of interest to the general public? Which of these might have financial returns, directly or indirectly, to the company? Focus on those that would have the greatest impact. As an example, GM has facilities in a third world country, a campaign to provide clean water to the surrounding communities would make headlines and win approval of environmental organizations. Toyota did just that. An assembly plant I visited on my GM-sponsored trip to Japan had a wastewater treatment system complying with the laws, but when asked by a small town below the plant for clean water, carbon adsorption filtration was added and clean water was sent to the town.

In the same way, the highly respected Cummins Company based in Indiana, has a reputation of taking care of their workers and their community, funding libraries, fire stations, and other public service facilities. The motto on their website is "Cummins serves and improves the world in which we live." This type of reaching out engenders good will toward companies.

**COOPERATE WITH RATHER THAN FIGHT THE EPA.**

One component of caring about communities is attempting to work in cooperation with the agencies that are expressing the public interest in these areas. As I have shown, GM spent years and millions of dollars to fight the regulating agencies rather than work with them. Regulatory resistance was standard procedure at General Motors. It was ingrained in the psyche of the corporation. But I personally saw the payoff from exceptions to this mindset. There were two examples given in Chapter Thirteen that prove that GM would benefit from reaching out to the regulatory agencies. A third example is found in a change in compliance attitudes that occurred at Saturn.

I was called to appear at President LeFauve's Saturn conference room after being accused by his staff for not being willing to use my influence to seek relief from the New Source Performance Standard.

Bob Potaker, the head of paint development, started the meeting by stating to LeFauve and his vice president that the Japanese plant near the Saturn site in Tennessee did not have to meet with the same emission standards.

They asked, "So why should Saturn?"

My reply was that the statement was a lie because no one was exempt from the federal law.

LeFauve, looking at Potaker, asked, "Is that true?"

He nodded yes. LeFauve's people had been lying to him.

I told LeFauve that Saturn would meet tough competition and to draw customers it needed the best paint finish in the world. Without pollution control devices, Saturn would be limited to selecting a paint that would satisfy regulations rather than customers.

When I got back to my office, LeFauve had Jim Cubbin, Saturn's attorney, call me to say what a good job I did and that he admired me for speaking the truth. LeFauve once told me that he had to take orders from GMAD but my words emboldened him to do what was best for Saturn. I called John Calcagni, EPA Durham, and told him that Saturn was building at great risk and would use waterborne paints and new technologies that may or may not work. I also said that the EPA must show that they can work with the American industry. John said he would give Saturn three years to comply after start-up and I thanked him. I went down the hall to the office of Bob Hoffman, who handled the Saturn permit, and told him I had just gotten five years for Saturn. And Saturn did get the five years.

Time and again cooperation saved the company money. GM would have been in bankruptcy much sooner had it not been for the Clean Air Act, the Clean Water Act and the Resource and Recovery Act. These regulations forced our engineers to analyze every operation for inefficiencies and waste. The result was not just pollution reduction but tremendous cost savings and better management of the production systems. In addition to reaching out to customers and the general public, reaching out in cooperation to the regulatory agencies can and will produce significant savings over GM's traditional method of fighting tooth and nail over every requirement.

## Reach Back

General Motors is a corporation with a rich history and a tremendous economy of scale. One of the wisest moves that GM can make is to look for ways to use what it has to reach back into the resources of the existing corporation and make the best use of what is already there—whether that be lessons from history, the experience and knowledge of those already on its payroll, or the billions of dollars' worth of buildings and equipment that it already owns. Maximizing utilization is a key to future financial health.

**Utilize the best lessons from the corporation's history.**

As I expressed at the beginning of this chapter, there are many lessons to be learned from GM's past. We should not be resigned to repeating the mistakes of the past, nor should we forget the leadership and business principles that initially built General

Motors. Amongst his many strengths, Alfred Sloan was famous for encouraging discussion and debate prior to a final decision. Debate was squelched from the 1970s onward in the culture of "team play" and this lack of thorough and honest analysis has been extremely costly in a multitude of ways.

The leadership of GM should apply the lessons of the past in the area of innovation. As noted in the overview of the corporation's history, GM led the way, introducing advances in independent front wheel suspension, automatic transmissions, and was the first to offer power steering, to name just a few examples. A downside of outsourcing component manufacturing and paint production is loss of control of research and development. A supplier will offer its innovations equally to the entire industry. GM needs to look at its core components and continually ask how they can be improved. The corporate structure needs to foster advances, not squelch them.

**Utilize the human resources already in place.**

Businesses should advance into leadership positions those employees that have expertise in the core business and have shown an ability to learn from their experiences. A corporation of GM's magnitude is full of knowledgeable engineers, managers, and laborers. Instead of promoting the "team player" and the young employee with "high potential," GM needs to utilize the knowledge and experience of those within its ranks.

This is the first of several areas where the Japanese companies have been more successful. Rather than dismissing them or disdaining them, every executive should take the opportunity to observe and learn. The Japanese have been much more effective in utilizing the full potential of human resources, and as in many of these categories for improvement, they can serve as a model.

In 1982, the GM Fremont plant was closed for a number of environmental and other business reasons, including labor problems. A year later, a joint venture with the Toyota Motor Corporation was formed, called New United Motor Manufacturing, Inc., or NUMMI, in order for the Japanese to use their methods to produce a small Nova Chevrolet car. Ideally, GM was to learn the secrets of the Japanese production methods and use the plant as an incubator for future top management by sending a dozen engineers to observe the plant's operations over a period of time. It was a good plan.

However, instead of selecting the most knowledgeable engineers to study the various types of assembly operations, the assignment was given to friends and relatives of GM executives (past and present) along with a word of encouragement: "This is Roger Smith's project and a great career move." What DeLorean labeled as "management by crony" ran the show.

I personally visited the Fremont plant and made several interesting observations. I found a high level of internal communication that fostered continuous improvement at every level. Before work began, the engineers and plant floor personnel met every morning to plan that day's plant improvements. A few of their improvements did make their way to other GM plants, but the communication system that fostered an atmosphere of continuous improvement was never replicated. As far as GM being an incubator for developing future top management talent, that didn't happen! The saddest part of the Fremont story is that the Japanese, relatively new in the business, took our plant and our labor force and produced a higher quality, lower cost, more fuel-efficient, and better selling small car than any of the big three U.S. manufacturers. Was their secret *not* being in operation for nearly one hundred years? The big surprise to the Americans was that the Japanese manufacturing technology was actually far less advanced than GM's. Their secret lay in the treatment of people, rather than in technology. They systematically harnessed the brainpower of the workers to improve plant efficiency. It was an approach not considered by GM.

Every business needs to give management positions to individuals who are passionate, devoted, and supremely qualified for the job. Consider if they understand the foundation of the business—in this case—automotive engineering. Consider if they have a track record of getting the job done and done well. Nepotism and the old "team player" promotion system should be avoided. Instead of personal gain, all decisions should be weighed by what is best for the company and its long-term bottom line.

This would be in contrast to Roger Smith's philosophy of staffing for the Saturn project. The story of David Skivens is one of many that I have cited in which an individual with a history of failed projects is given more responsibility instead of less. Skivens led the charge on the Paint Plan and then at a time when the failings of the Plan were evident, he was appointed to head the construction of the Saturn plant. He wasn't a civil engineer and as time would show, not just the Saturn product, but the plant itself was an investment disaster.

Place the very best person in every job and give them the leadership of supervisors that have proven track records and are fanatical about their particular part of the business operation. Give employees reason to be proud of their part of the corporation. At the highest level, insist that at least two members of the board come with direct experience and documented success in the core business function, which in this case is manufacturing.

**UTILIZE THE EXISTING PHYSICAL ASSETS TO THEIR MAXIMUM.**

In the same way that a business should utilize the knowledge and experience of its best employees, GM needs to reach back into its wealth of capital investment in buildings and equipment and maximize utilization of the physical assets that it already owns. There was a false hope in technology and "everything new" during the Roger Smith era. There was an assumption amongst American businesses that the Japanese and Chinese competitors were gaining market share because they were using robots and other forms of cutting-edge technology. But this was not their secret.

Since GM is the oldest and biggest in the business, companies just starting in the automotive business should be flocking to Detroit to learn how it is done, right? To the contrary, the November 20, 2006, *New York Times* reported, "GM executives from all over the world have been traveling to...an obscure city in the southernmost China, to learn the secrets of low-cost production."[31] They discovered that the Chinese company, Wuling Motors, used existing buildings, if possible. They applied new processes and technologies only when they were proven to work and be cost effective. They only automated a worker's station or replaced a worker with a robot when it was proven to be productive and profitable. "The biggest answer seems to lie in Wuling's ability to minimize the investment cost of everything from buildings to tooling to engineering to automation, Mr. Wale, the president of GM China, said. 'If they have to do some mechanization,' he said, 'they find the very cheapest way.'"[32]

Sounds like common sense. At the time, China had been in the automotive business for a mere twenty years. Why would GM, with nearly one hundred years of history need to seek out China's manufacturing systems as a model?

Even with new models that are garnering better reviews, GMAD is still in need of a major overhaul. How the assembly plants are used, i.e. "plant utilization," is the most important factor in saving the new GM and making it financially competitive. Maximizing the return on investment on these billion dollar facilities is the key to future profitability. GMAD's assembly plants must strive to produce more vehicles on the same line, operate twenty-four hours a day, at least six days a week, and also produce other companies' vehicles, as was once done by Toyota for GM.

In 2014, GM had 11 active truck and car plants. Four of these produced roughly 25,000 to 31,000 units per month. At this level the plants produced a solid profit. Notably, two of these plants, Lordstown, Ohio (built in 1966), and Arlington, Texas (built in 1953), are the oldest—yet among the top producers. Then there are four plants producing from 12,000 to 17,000 units per month. These plants are in the black, but clearly not as profitable. Finally, there are three plants, including what used to be the Saturn plant, which are operating in the red. According to the GM website, they produce between 4,000 to 6,000 units and sometimes less. In the case of Saturn, underutilization is preferable to the rise of community ire that would result from a complete closure.

What if a plant is producing a vehicle that has weak sales figures? How do you increase plant utilization without creating excess inventory? To achieve this, the assembly plant must be capable of producing more than one model and/or quickly changing from one model to another. The new Ford truck plant can produce as many as seven different models. Other manufacturers have kept model changeover time to a minimum.

Consider the example of Honda and now Ford in efficient retooling. Historically, GM has been inefficient in this area, allowing plants to sit idle for months while they are retooled. In this age of instant communication and rapidly changing markets, GM must do much better in the art of flexible manufacturing. As far back as

1990, Honda had the capability to switch from one model to another while the production line kept moving. The secret is in knowing to what degree of articulation, or span of movement, the robotic arms, automatic welders, and other production devices can travel and still perform their function. Designers, knowing these operating boundaries, can provide a facelift to an existing vehicle or plan for a completely new model, without causing a major modification to the assembly line equipment.

Chrysler has seen tremendous success in emphasizing utilization at its Jefferson Avenue plant in downtown Detroit. According to the *New York Times*, "The Jefferson North plant generates an estimated $2 billion in annual profits and is a huge contributor to Chrysler's health."[33] The plant produces 300,000 vehicles annually, triple the amount produced in 2009 when Chrysler filed for bankruptcy. Vlasic goes on to report that according to Kristin Dziczek, an analyst at the Center for Automotive Research, "You have a hot-selling, high profit vehicle, a flexible labor agreement, and a facility that the company has invested in instead of abandoning."[34] In other words, Chrysler has made maximum use of the investments it had made in the past.

GM may have yet to learn the secret of utilization but it appears in 2013 that the Ford Motor Company has taken note. Week after week, the Detroit papers reported that with sales volumes high, Ford refused to succumb to the pressure to construct a new plant. Instead they were doing all they could to utilize their existing facilities to the maximum by adding round the clock shifts and making use of every inch of space.

Utilization of history lessons, improved utilization of the human resources, and maximized utilization of physical assets can together strengthen any business's bottom line.

## Reach Down

Whether the issue is planning, implementing, or correcting, there is example after example of scenarios within GM where qualified oversight was lacking. In the early chapters, I documented the lack of engineering knowledge on the corporate board and in much of the upper management. Foolish plans were proposed and then approved for implementation. Money was thrown at projects with little or no accountability for results and return on investment. It has been a refrain of mine for decades that GM needs to reach down into the layers of the corporation with meaningful analysis and review.

I have in my files a copy of a memo I wrote and circulated in January of 1988 proposing the formation of a "Painting Technologies Steering Committee." The committee described is a prototype of what was and is needed within any manufacturing company. Such a committee would be charged with assessing current systems for cost efficiency and developing means of information exchange with accessible storage of key technical information. They would establish R&D priorities and keep pushing technologies toward the highest quality at the lowest cost. Sounds logical and reasonable doesn't it? Robert Frosch, the vice president for GM research, liked my recommendations but couldn't get GMAD's approval.

A successful company needs to have established channels of supervision to keep its business on track and to thwart the damaging tendencies toward decisions motivated by pride and self-interest. Accountability needs to extend beyond the area of finances.

For years, our Environmental Activities Staff only audited the U.S. plants every three or four years for pollution compliance with little playback except to tell stockholders that we care. I have always proposed that a compliance form be sent to all plants—annually to plants with serious pollution sources, biannually to others—requiring the plant manager to certify compliance with all the rules. If the plant manager lies, causing great embarrassment to the corporation and costs for violations, he should be fired immediately. This only has to happen once and all plant managers get the message that pollution compliance is important and dishonesty will no longer be tolerated. Recently, there was an overseas plant with a pollution problem; in response, the current CEO instituted a requirement that all environmental engineers worldwide report directly to an executive in Detroit. This added a layer of reporting but it failed to create a stronger incentive for avoiding future pollution problems.

It was my direct observation that GMAD did not show a careful engineering cost evaluation of assembly plant projects, nor demonstrate evidence of continuous manufacturing improvements based on a constant measurement of performance at the plants. This was evidence of "resting instead of reaching."

Efficiency and cost/benefit need to be measured against a goal and reported to management. Constant manufacturing improvements can only come from measuring the performance of all the plants to a set of standards of operations. For example, at GM, an assembly plant operational office should be established that is independent of GMAD. It should be staffed with experts and specialists, many from outside the corporation, who have leadership experience in assembly and paint operations. It would have the authority to order and collect performance data and to make recommendations to the plants for improving utilization and making upgrades on a priority basis. An annual 5–10% improvement in efficiency in plant performance would be mandated. In all areas of operation, from plant construction to fabrication, to assembly and paint operations, any innovation should be funneled to a "Continuous Improvement Group" so it can be evaluated for application in other plants.

Financial and marketing staffs would undergo a complete overhaul, with these staffs weighing in heavily on project decisions. They would be held accountable for projects approved as well as for decisions based on their sales forecasts.

## REACH FORWARD

Once the foundation of reaching out to the consumers and agencies, reaching back to make wise use of historical, human and capital resources, and reaching down with meaningful management oversight has been laid, *then* it is time to reach forward into the future!

**VALUE RESEARCH.**

No business can afford to rest on its laurels. Competitors will pass you by and take over your market share. Innovation has always been a hallmark of a free market economy and it always will be. IBM for example, has always been a leader in advance technology with signs all over the company spaces saying "Think!" GM needs to, and is capable of, returning to its position as an innovator, developing and testing potential advances in-house. To accomplish this will require some intentional reordering of priorities.

As an example of what could be done in several key business areas, GM must create an environment to improve paint technologies rather than wait to follow the lead of others. Chrysler was applying the new "wet on wet" paint application system and saving money while GM was taking a wait and see attitude.

To do this a $10 million award should be established for the development of a less expensive method for the painting of vehicles. The current cost of the paint spray booths, the robotic applicators, and the paint itself has topped the cost of the remaining operations of an assembly plant, making cost-reduction in this area a worthwhile priority. In the expanding global economy, paint pigments, resins, additives, and solvents that go into making the final product are increasingly scarce and expensive. Other finishes, methods of applying coatings, and less expensive materials must be discovered, along with new abatement and application techniques.

The Paint Plan promised GM management that their products would have a world-class finish. With this in mind, the following approaches would seem logical. They were not applied then, but should be in the future.

- Research must not be limited in any way.
- Always keep the wants and needs of the customer in mind.
- Keep the paint superintendent in the loop of communication.
- Engineers should be free to talk to any company and any branch of any agency to discover what paints, abatement controls, and application technologies are being used or developed world-wide (with special interest in paints that are solvent-free).
- Engineers must evaluate these emerging technologies, abatement systems, and paints and provide management, sales, marketing, etc., with a number of options so that the best system for the company is chosen.

The paint quality of a vehicle is a factor in sales and customer satisfaction. A concentration of research efforts should address the issue of quality as well as cost. In some colors, waterborne paints do provide a satisfactory finish without the addition of stack emission controls. But not all colors are to General Motors' previous standards for lacquer finish. Recently, a new black Cadillac was observed to have a high degree of orange peel, a finish considered unacceptable when lacquer paints were used.

**FIRST LOOK FOR SOLUTIONS.**

Instead of settling for the status quo and ignoring problems, GM needs a new manufacturing culture and should hire the best process engineers in the world to concentrate on making GM's assembly plants the most utilized and efficient in the world. Today they are widely known to be the least efficient in the industry.

To reduce the number and cost of environmental problems, process manufacturing engineers should be held accountable for all manner of problems generated by a process or production system, and not require separate environmental engineers to look over their shoulder for environmental issues.

The bottom line is that the recent GM manufacturing culture has never concentrated on making the assembly plant operations the best in the world. An extremely accurate statistical analysis of every assembly plant's productivity must be instituted and modernization projects measured for their benefit to the corporation before a similar assembly plant project is undertaken.

**ANTICIPATE AND RESOLVE PROBLEMS BEFORE REGULATIONS ARE ENFORCED.**

Proactivity was a practice that I observed in Japan. Because of our new relationship with Toyota, formed in 1983 as part of NUMMI, GMAD was offered a courtesy visit to Japan to learn how they respond to environmental problems. Because of short notice, there wasn't time to arrange a visit with their environmental control agencies. Nevertheless, the trip proved extremely useful. As the only executive, I acted as company spokesperson.

We were unexpectedly greeted by Toyota's president in the conference room and introduced to our tour guide to the plant—the person who developed activated carbon into a fabric filter to control spray booth emissions. The trip got more exciting days later. The group was transported by the cleanest taxi in the world, the smoothest and cleanest on-time street cars, and the famous bullet train traveling over 120 miles per hour. I also noted that trucks leaving construction sites had to be washed to avoid throwing stones and dirt on streets and cars.

But it was the environmental problem solving that impressed me the most. For example, we visited a stamping press metal plant that had annoyed a senior housing project nearby with its constant pounding all day. So the entire plant, walls and roof were soundproofed with extremely thick acoustical panels, not wood or brick walls like used in the U.S. to reduce highway traffic noise. Likewise, the road back to the airport had acoustical panels built over the eight-lane highway extending probably twenty feet above with a ten-foot gap in the middle to allow air pollution to escape. And in the area of special interest to our group, I noticed that paint systems causing community odor problems were controlled as soon as possible.

When I inquired, *why do you solve problems quickly?* The answer was *to avoid the government passing laws forcing everyone to do the same.* Proactivity rather than resistant reactivity was their secret. Japanese companies, to the extent possible, solve problems without getting the government involved. U.S. industry, in contrast, seeks an even playing field for all the competitors, meaning government must force everyone to comply. Their actions (or inactions) *bring* government into their business. The two cultures are very different, and this difference in philosophy toward government has been costly for GM.

In closing, my memoir was written for personal satisfaction and to present the facts on how General Motors works internally. I'm sure some events described will embarrass individuals mentioned; a few may be hurt. But not to express my true feeling or recount actual statements and events would be a serious mistake. So I expect to be attacked by some who will disagree with my observations or conclusions, but years of evidence are stacked in my favor. I share my story to reveal the detrimental hidden culture with hopes for leadership that can effect a positive and lasting change.

So let's get at it.

# Epilogue

I BEGAN THE PROCESS OF PUTTING MY LONGSTANDING CONCERNS about GM's culture into a book format in 2010. At that time, as I have noted, my family expressed concern about the repercussions of going public with my critique. As this goes to print in 2015, the headlines are full of public concerns over GM's internal culture. The *very same issues* that I have recorded are finally getting the level of attention that I hope will precipitate a lasting change. Failures to report and correct safety issues related to an ignition-switch flaw have resulted in massive recalls, liability lawsuits, and worst of all, the loss of life.

In congressional hearings regarding the Cobalt ignition-switch defects, GM's current Chief Executive Officer Mary Barra said, "Yes, she was a 33-year veteran of the company, but she was as mystified by the actions of the old General Motors as anyone."[35]

Her statement confirmed my opinion that she and others at the board level and in top management haven't the slightest idea what the company culture is. Apparently, she was unaware of GM's long standing culture of resistance to safety regulation and implementation and its disregard for the customer. A *New York Times* article of September 8, 2014, stated that the current board chairman, Theodore Solso, was "shocked" and "stunned" by a 315-page report that "uncovered years of errors, missteps and deception among engineers, lawyers and product specialists." It was no shock to me.

All of the concerns now coming to light had been chronicled in this book. I had observed the same attitude of resistance in the area of environmental pollutants. I observed first hand, over a period of several decades, the ignorance of the GM board, the lack of oversight and reporting of problems. Financial rewards went to those vice presidents, engineers and lawyers who both told their bosses only good news and were successful in delaying the implementation of vehicle safety requirements. Those who forestalled control expenditures received the highest salaries, even when, like in the Cobalt recall case, the cheaper parts later cost the company millions more.

The Cobalt ignition-switch locator plunger and back-up spring should have been a mere 1/8 inch longer. At a cost of a few pennies per switch, the slippage of the plunger could have been prevented. Now, billions will be spent on recalls and lives lost cannot be replaced.

There was a report of an ignition-switch accidently changing to the off position as far back as 2001, with sporadic reports each year until 2004, finally prompting an engineering design change. Congress and the public want to know why action wasn't taken sooner and why management wasn't told of the problem. Most likely, the engineers, who did not react to these reports, hoped that an incidence of an ignition-switch accidently changing position would be rare and would not cause deaths. When making decisions, every automotive engineer considers, at least subconsciously, how to save pennies. Cents per part, times millions of cars produced, means millions to the bottom line. Not telling management about problems is the accepted procedure when there is no directive to do otherwise.

Most auto companies have had similar "don't tell" employee problems. Only Ford has had a better track record dating back to Henry Ford II who, as I stated earlier, was quoted to say about a death caused by a fan blade, "God, save me from the fellow that wants to save me."

In 2006, when Alan Mulally became Ford Motor Company CEO, he discovered information critical to the business that was not being shared. It was a cultural problem that was corrected by mandatory weekly meetings of senior executives, during which they had to report on progress or lack of progress in solving problems. At Ford, Mulally actually heaps praise on executives willing to disclose sensitive information about company problems in his efforts to change Ford's culture. At GM, problems with the Paint Plan were presented to management time and time again, with no change in the program's direction throughout the 10 years.

The three CEOs since the company came out of bankruptcy, including Barra, have no idea what this "culture" thing is. I distinctly remember a conversation I once had with a friend, who had spoken to the secretary of GM's Ed Whitacare, the first of the three CEOs since the bankruptcy. The secretary had confessed to my friend that Whitacare had asked her, "What's this culture stuff I'm supposed to change?" He had no clue. To fully understand the company culture one must have worked in the midst of it. I hope my book has answered his question.

There is still time for GM to change its culture to act on reports of problems, and take the lead on safety, quality, and fuel economy. GM's CEO Barra at the hearings also said, "We will now focus on the goal of becoming an industry leader"—although not *the* leader—"in safety!"[36] Recently GM spokesman Greg Martin said, "Our first, guiding principle is to put the customer at the heart of all we do."[37] Never was that my observation in 35 years at GM, but perhaps it can become that.

The company culture can be best modified by concentrating on new policies and directions specifically targeted at employees responsible for 1) regulatory government proposals and 2) considering the public's interest in everything the company does. The product engineers are developing great vehicles and improving quality but need additional guidance in these two historically problematic areas. A major cultural change in attitudes toward government regulatory proposals and public interest can eliminate ignition-switch type problems.

The CEO Barra should clearly state:

- This is what we will do
- This is what we will not do
- This is what is rewarded
- This is what is punished

As this work goes to press, I have a feeling of satisfaction after years of writing with little expectation of publication. The insights revealed in this book are needed. Books about the company's bankruptcy and months of news articles about the current Cobalt ignition-switch all blame the company's culture. Yet those pointing to the culture fail to clearly express what it is, how it was created, who or what staff are most influenced by it, or what should be done to change it. My satisfaction comes from the knowledge that my book addresses these questions.

# Endnotes

## Chapter One

1. Maryann Keller, *Rude Awakening* (New York: Harper Perennial, 1990), 196.
2. Steve Rattner, "The Auto Bailout: How We Did It," *Fortune*, November 9, 2009, 71.

## Chapter Three

3. Burton W. Folsom, "Billy Durant and the Founding of General Motors," accessed January 3, 2013, www.mackinac.org/651.
4. J. Patrick Wright, *On a Clear Day You Can See General Motors* (New York: Avon Books, 1980), 252.
5. www.gm.com/company/historyAndHeritage/revolution.html accessed August 27, 2014.

## Chapter Four

6. Wright, *On a Clear Day You Can See General Motors*, 199.
7. *The Economist*, February 16, 2010.
8. Alfred Sloan, *My Years at General Motors* (New York: Doubleday, 1963), 44.
9. Isikoff, Michael with Stephan Kuffner "A $16 Billion Problem," *Newsweek*, August 4, 2009, 27.

## Chapter Five

10. Wright, On *a Clear Day You Can See General Motors*, 53.
11. Keller, *Rude Awakening*, 95.
12. Ibid, 94.
13. Bernadine Williams, "Outsider Elmer Johnson Shook Up GM," *Automotive News*, March 24, 2008.
14. Paul Ingrassia and Joseph White, *Comeback* (New York: Touchstone, 1994), 167.

## Chapter Nine

15. Keith Bradsher, "G.M. Plans Shift to Small Cars for the Emerging World" *New York Times*, November 20, 2006.
16. General Motors Corporation, "Study to Determine Capabilities to Meet Federal EPA Guidelines for Volatile Organic Compound Emissions" (Detroit: November 1, 1978), 11.
17. Ibid, 11.
18. Ibid, 15.
19. "A History of Powder Coatings" last modified February 19, 2004, www.pcimag.com/articles/a-history-of-powder-coatings.

## Chapter Ten

20. G.R. Maher, "Air Pollution Regulation of Nonvehicular, Organic-Solvent Emissions by Los Angles Rule 66," *Journal of the American Oil Chemists' Society*, August 1967, A340.
21. Ibid.

## Chapter Eleven

22. Wright, *On a Clear Day You Can See General Motors*, 199.
23. Ibid, 252.

## Chapter Thirteen

24. "Tricky Technology" *Wall Street Journal*, May 13, 1986, 1.
25. Wright, *On a Clear Day You Can See General Motors*, 119.
26. General Motors Corp v. United States, 496 U.S. 530 (1990).
27. EPA Enforcement Accomplishments Report, FY 1990, Office of Enforcement, February 1991.
28. EPA Enforcement Accomplishments Report, FY 1991, Office of Enforcement, April 1992, 4-3.

## Chapter Fourteen

29. Keller, *Rude Awakening*, 227.

## Chapter Fifteen

30. Charles Duhigg, *The Power of Habit* (New York: Random House, 2012).
31. Bradsher, "G.M. Plans Shift to Small Cars for Emerging World."
32. Ibid.
33. Bill Vlasic, "Last Car Plant Brings Detroit Both Hope and Cash," *The New York Times*, July 16, 2013, A1.
34. Ibid.

## Epilogue

35. Doron Levin, "New GM: Same as it Ever Was?" *Fortune Magazine*, May 2014, 64.
36. Matthew L. Wald and Daniel Ivory, "GM is Fined over Safety and Called a Lawbreaker," *The New York Times*, 17 May 2014, b2.
37. Quoted from an email written by Martin in "GM's Supplier-Squeezing Days Gave Birth to Flawed Models" by Keith Naughton, David Welch, Jeff Green and Mina Kimes, *Bloomberg*, March 21, 2014, www.bloomberg.com.

# About the Author

Nicholas Kachman retired as the assistant director of Plant Environment after thirty-five years of employment with General Motors Corporation. His accomplishments include initiating the GM Corporate Noise Abatement Program and directing General Motor's Air Pollution Control Program. Kachman is nationally recognized for his accomplishments in the fields of environmental studies and engineering and also travelled extensively overseas on behalf of GM. A frequent university guest speaker and lecturer, he was appointed to numerous environmental advisory committees by the American Lung Association and by three different Michigan governors.

Kachman served for ten years on the Michigan Air Pollution Control Commission with one year as chairman. In recommending him for that commission, Howard Tanner, the director of Michigan's Department of Natural Resources wrote, "Mr. Kachman is not only extremely knowledgeable in the field of air pollution but also an expert on environmental protection in general...a person who all industry would accept as a commission representative with unquestionable integrity, outstanding technical ability and insightful understanding of industrial and business perspectives."

Nick lives in Michigan with his wife, Isabel. They have five grown children and several grandchildren and great-grandchildren. In his retirement, he enjoys fishing for salmon and hunting turkey and elk, donating much of the elk meat to rural food pantries. He also enjoys travelling internationally with his son, exploring world cultures. A native of Detroit, he graduated from Wayne State University with a BSME.

CPSIA information can be obtained
at www.ICGtesting.com
Printed in the USA
FFOW01n0937190615
14391FF

9 780990 965343